What Functional Managers Need to Know about PROJECT MANAGEMENT

Harold Kerzner, Ph.D.

Frank Saladis, PMP

WILEY

John Wiley & Sons, Inc.

INTERNATIONAL
Institute for Learning, Inc.

Photo credits—2.2, 2.3, 2.8, 2.9, 4.2, 4.7, 4.8, 4.11, 4.13, 4.18, 4.19, 4.20, 6.7, 6.8: PhotoDisc/Getty Images; 2.4, 2.5, 2.7, 4.5, 4.10, 4.12, 4.16, 4.17, 6.4: Digital Vision; 2.1, 2.6, 3.1, 6.6: Purestock; 2.10, 4.9, 6.3: Artville/Getty Images; 2.11, 4.14: Corbis Digital Stock; 2.12, 6.5: Imagestate

"PMI", the PMI logo, "OPM3", "PMP", "PMBOK" are registered marks of Project Management Institute, Inc. For a comprehensive list of PMI marks, contact the PMI Legal Department.

Library of Congress Cataloging-in-Publication Data:
Kerzner, Harold.
    What functional managers need to know about project management / Harold Kerzner, Frank Saladis.
        p. cm.—(The IIL/Wiley series in project management)
    Includes index.
    ISBN 978-0-470-52547-0 (cloth)
        1. Project management. I. Saladis, Frank P. II. Title.
    HD69.P75K4973 2009
    658.4'04--dc22
                                                                    2009018591

Printed in the United States of America

V10011821_070219

CONTENTS

PREFACE

When project management first began, the only industries that readily embraced project management as a way of doing business were aerospace, defense, and heavy construction. These industries were identified as project-driven industries, where each project had a profit target. The prime objective of project management was to generate profits, and the project managers had the responsibility for profit and loss. The survival of the company rested in the hands of the project managers.

Project managers were viewed as managing profit centers, and functional manager groups were looked upon as cost centers. The role of the functional manager was basically to assign resources to projects and to keep their costs as low as possible. If the project was successful, then the project manager could expect to receive a bonus. If the project failed, blame was placed on both the project manager and the functional managers. Functional managers were treated with indifference and often received neither bonuses nor credit for doing their job well.

Functional managers were not required to understand project management. Their role was to assign resources to projects and often relied on the project managers to provide daily direction to the resources. The reason for this was that project managers at that time were, in almost all cases, engineers with advanced degrees and they possessed a strong knowledge and command of technology, often a greater technical knowledge than the functional managers. Functional managers would basically relinquish any control over the resources once the resources were assigned to the projects.

As project management matured and the projects became more sophisticated, it became extremely difficult for project managers to maintain their technical expertise and continue to possess a command of technology. Many were no longer considered to be technical experts. Most project managers today have an understanding of technology rather than a command of technology. The technical expertise resides in the functional areas. As a result, the accountability for the success of the project is now viewed by many executives and project sponsors as shared accountability between the project manager and all participating line or functional managers. With shared accountability, the line managers must now develop a good understanding of project management, which is why more line managers are now seeking project management certifications and credentials such as the Project Management Institute's PMP® (Project Management Professional) and CAPM® (Certified Associate Project Manager). Today, project managers are expected to focus on and manage project deliverables rather than people. Management of the assigned resources has become a line function.

Today, the technical knowledge repository of most companies resides in the functional areas. When resources are assigned to a project, the resources continue to receive technical direction from their functional managers. Functional managers are now an integral part of project management and share in the success and failure of each project. Project management is now viewed as a discipline of team leadership and team accountability.

As project management continues to evolve and mature, the relationship between the project manager and functional managers continues to mature and is getting stronger. They must work together, understand each other's priorities and problems, and resolve issues jointly. When a functional manager encounters a problem when assigning resources, the functional manager goes directly to the project manager for assistance and contingency planning. When a project manager has a resource-related or technical problem, they go to the functional managers for assistance with the identification of alternatives. Senior management may be called upon to assist in problem resolution only as a last resort if the project and functional managers cannot reach an agreement. The success of project management may very well rest upon how well the project manager and functional managers work together.

The role of the functional manager has changed significantly. Functional managers now have the power and influence to drive any project to success or failure by the way they provide support for the project. Therefore, a positive relationship between project manager and functional manager is essential. This need for a strong project manager/functional manager relationship has become apparent in the implementation of the majority of today's projects, and senior management has finally realized the importance of functional management in making project management succeed.

HAROLD KERZNER
FRANK SALADIS
INTERNATIONAL INSTITUTE FOR LEARNING, INC. 2009

ACKNOWLEDGMENTS

Some of the material in this book has been either extracted or adapted from Harold Kerzner's *Project Management: A Systems Approach to Planning, Scheduling, and Controlling*, 10th edition; *Advanced Project Management: Best Practices on Implementation*, 2nd edition; *Strategic Planning for Project Management Using a Project Management Maturity Model*; *Project Management Best Practices: Achieving Global Excellence*, 1st edition (all published by John Wiley & Sons, Inc.).

Reproduced by permission of Harold Kerzner and John Wiley & Sons, Inc.

We would like to sincerely thank the dedicated people assigned to this project, especially the International Institute for Learning, Inc. (IIL) staff and John Wiley & Sons, Inc. staff for their patience, professionalism, and guidance during the development of this book.

We would also like to thank E. LaVerne Johnson, Founder, President & CEO, IIL, for her vision and continued support of the project management profession, Judith W. Umlas, Senior Vice President, Learning Innovations, IIL, and John Kenneth White, MA, PMP, Senior Consultant, IIL for their diligence and valuable insight.

In addition, we would like to acknowledge the many project managers whose ideas, thoughts, and observations inspired us to initiate this project.

—HAROLD KERZNER, PH.D., AND FRANK SALADIS, PMP

INTERNATIONAL INSTITUTE FOR LEARNING, INC. (IIL)

International Institute for Learning, Inc. (IIL) specializes in professional training and comprehensive consulting services that improve the effectiveness and productivity of individuals and organizations.

As a recognized global leader, IIL offers comprehensive learning solutions in hard and soft skills for individuals, as well as training in enterprise-wide Project, Program, and Portfolio Management; PRINCE2®,* Lean Six Sigma; Microsoft® Office Project and Project Server,** and Business Analysis.

After you have completed *What Functional Managers Need to Know about Project Management,* IIL invites you to explore our supplementary course offerings. Through an interactive, instructor-led environment, these virtual courses will provide you with even more tools and skills for delivering the value that your customers and stakeholders have come to expect.

For more information, visit http://www.iil.com or call +1-212-758-0177.

---

*PRINCE2® is a trademark of the Office of Government Commerce in the United Kingdom and other countries.
**Microsoft Office Project and Microsoft Office Project Server are registered trademarks of the Microsoft Corporation.

1

PROJECT MANAGEMENT PRINCIPLES

PROJECT MANAGEMENT HUMOR

Project management is the art of creating the illusion that any outcome is the result of a series of predetermined, deliberate acts when, in fact, it was dumb luck!

Some people are under the impression that project success is accomplished by chance and luck. Nothing could be further from the truth. Most people will agree that project success is accomplished through a structured process of project initiation, planning, execution, monitoring and control, and finally closure.

Some companies rely heavily on an organized and consistent project management methodology to accomplish their goals. Some methodologies are based on policies and procedures, whereas others are developed around forms, guidelines, templates, and checklists.

Project management is an attempt to get nonroutine work to flow multidirectionally through the company, usually horizontally, rather than in a vertical, sometimes bureaucratic manner. To accomplish this multidirectional work flow, a project management methodology is required. One of the purposes of this structured methodology is to facilitate the job of integrating the work across various functional units to meet project objectives.

When projects reach completion or closure, the project team is debriefed in order to capture lesson learned and best practices that may be beneficial to the organization and for use on future projects. In most cases, the best practices that are discovered are used to improve how the project and functional managers interface and to increase efficiency in the use of organizational resources.

PROJECT MANAGEMENT

PROJECT PLANNING

- Definition of work requirements
- Definition of quantity and quality of work
- Definition of resources needed

PROJECT MONITORING AND CONTROL

- Tracking progress
- Comparing actual outcome to predicted outcome
- Analyzing impact
- Making adjustments

The *Guide to the Project Management Body of Knowledge* (PMBOK®
Guide) identifies five domain areas in which the project managers
must perform:

- Initiation—Defines and authorizes the project

- Planning—Defines and refines project objectives

- Execution—Integration of resources to meet objectives

- Monitoring and Control—Measuring progress and identifying
  variances

- Closure—Acceptance of project deliverables

The amount of time and effort that project managers must put
forth can vary based on the domain area. Many project managers are
not brought on board the project until the end of the initiation process. Executive management, marketing, and sales may take the lead
during project initiation.

Project managers and functional managers are heavily involved
in project work during planning, monitoring and control. During
project execution, much of the work is accomplished by the project
team and the functional managers. If the project team members
report directly to their specific functional departments, the project
manager's main contact with these resources may be during monitoring and control of project activities as tasks are executed.

During project closure, the project manager is expected to make
sure that all project documentation is complete and ready for the
archives. Some companies bring on board project closure experts to
shut down large projects.

PROJECT NECESSITIES

- Complete task definitions
- Resource requirement definitions (and possibly skill levels needed)
- Major timetable milestones
- Definition of end-item quality and reliability requirements
- The basis for performance measurement

Planning is often regarded as the most important activity for a project manager. The project manager must understand the following:

- *All of the tasks necessary to accomplish the deliverables.* Many times the project manager does not possess a command of technology and must rely upon the functional managers for clarification and identification of project components, activities, and their respective risks.

- *Functional skills needed to accomplish the work.* The functional managers may be in a better position than the project manager to identify the skill levels needed to complete project work.

- *Major milestones identified by the customer, whether an internal or external customer.* The functional managers must verify that they can meet the milestone dates. Functional manager commitment is essential.

- *Quality of the deliverables.* The functional managers must confirm that they can meet the customer's quality and specification requirements.

- *Performance measurement.* The functional managers and project manager must agree about how to measure project performance with reference to the work breakdown structure (WBS) and detailed activity lists developed by the project team. It is possible that the WBS may require some changes and updates to support the functional manager's tracking processes.

RESULTS OF GOOD PLANNING

What are the results of good project planning as seen through the eyes of the functional managers?

The following points define the results of good planning:

- Assurance that functional units will understand their total responsibilities toward achieving project needs.

- Assurance that many of the problems associated with the scheduling process and allocation of critical resources are identified and are addressed through risk management.

- Early identification of risks and issues that may jeopardize successful project completion and the corrective actions required to prevent or resolve problems.

- A plan has been established for the purpose of guidance, problem solving, and decision-making, which will allow functional managers to spend more time supervising their people rather than resolving conflicts and solving problems.

PROJECT CHARACTERISTICS

- Have a specific objective (which may be unique or one of a kind) to be completed within certain specifications

- Have defined start and end dates

- Have funding limits (if applicable)

- Have quality limits (if applicable)

- Consume human and nonhuman resources (i.e., money, people, equipment)

- Be multifunctional (cut across several functional lines)

B efore continuing on, we should provide a definition of a project. Projects are most often unique endeavors that have not been
attempted before and might never be attempted again. Projects have
specific start and end dates. In some cases, projects may be very similar or identical and repetitive in nature, but those situations would
be an exception rather than the norm. Because of the uniqueness of
projects and their associated activities, estimating the work required
to complete the project may be very difficult and the resulting estimates may not be very reliable. This may create a number of problems and challenges for the functional manager.

Projects have constraints or limitations. Typical constraints include
time frames with predetermined milestones, financial limitations,
and limitations regarding quality as identified in the specifications.
Another typical constraint may be the tolerance for risk and the amount
of risk that the project team or owner can accept. There may also be
limitations on the quality and skill levels of the resources needed to
accomplish the tasks.

Projects consume resources. Resources are defined as human—
people providing the labor and support—and nonhuman—equipment,
facilities, and money, for example.

Projects are also considered to be multifunctional, which means
that projects are integrated and cut across multiple functional areas
and business entities. One of the primary roles of the project manager
is to manage the integration of project activities.

THE TRIPLE CONSTRAINT

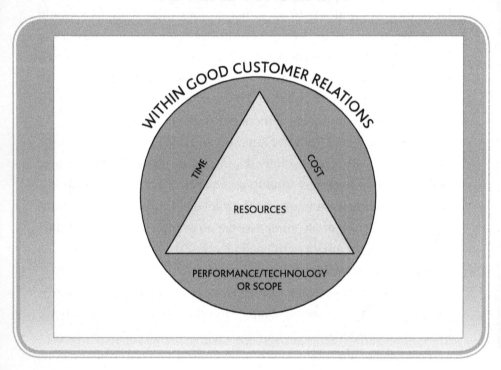

Project management is an attempt to improve efficiency and effectiveness in the use of resources by getting work to flow multidirectionally through an organization. Initially, this might seem easy to accomplish, but there are typically a number of constraints imposed on a project. The most common constraints are time, cost, and performance (also referred to as scope or quality) and are known as the triple constraint.

From an executive management perspective, the preceding illustration is the goal of project management, namely, meeting the triple constraints of time, cost, and performance while maintaining good customer relations. Unfortunately, because most projects have some unique characteristics, highly accurate estimates may not be possible and trade-offs among the triple constraint may be necessary. Executive management and functional management must be involved in almost all trade-off discussions to ensure that the final decision is made in the best interest of both the project and the company. Project managers may possess sufficient knowledge for some technical decision making, but may not have sufficient business or technical knowledge to adequately determine the best course of action to address interests of the company as well as the project.

RESOURCES

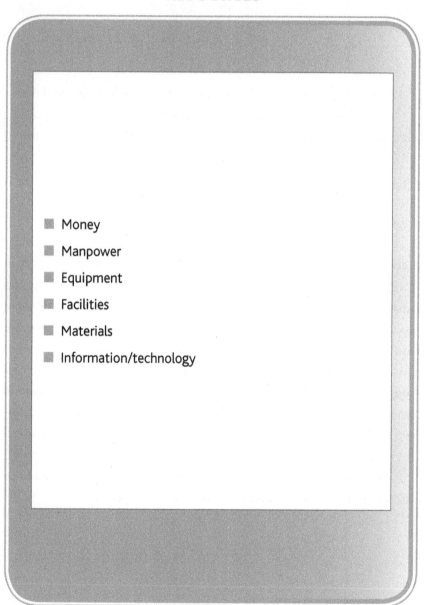

- Money
- Manpower
- Equipment
- Facilities
- Materials
- Information/technology

H ere are some of the typical resources that are used when executing projects. Assuming that the project manager and functional manager are separate roles assigned to different people, the resources are generally administratively under the control of the functional managers. The project managers must therefore negotiate with the functional managers for some degree of control over these resources. It is not uncommon for project managers to have minimal or no direct control over project resources and must rely heavily on the functional managers for resource-related issues. The resources may be in a solid line type of reporting relationship to their functional manager and dotted line or indirect reporting to the project manager.

Some people argue that project managers have direct control over all budgets associated with a project. The truth of the matter is that project managers have the right to open and close charge numbers or cost accounts for a project. But once the charge numbers are opened, the team members performing the work and their respective functional managers are actually in control of how the money is being spent as long as the charge number limits are not exceeded.

There is an exception, however. If the project work must be performed at a remote location where the employees are physically removed from their functional area, the project manager may actually have direct control of the resources. This is quite common on construction projects.

TYPES OF PROJECT RESOURCES

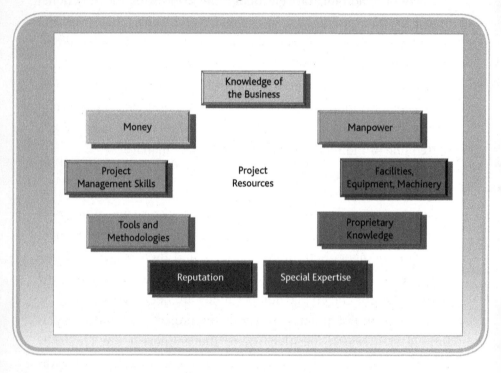

The illustration shows the various project resources that project managers may or may not have under their direct control. Some of these resources require additional comment.

- *Money.* As stated previously, once budgets are established and charge numbers are opened, project managers focus more on project monitoring of the budget rather than management of the budget. Once the charge numbers are opened, the performer or workers and their respective line managers control how the budgets for each work package will be used.

- *Resources.* Resources are usually "owned" by the functional managers and may be directly controlled by the functional managers for the duration of the project. Also, even though the employees are assigned to a project team, functional managers may not authorize them to make decisions without review and approval of the functional managers.

- *Business knowledge.* Project managers are expected to make business decisions as well as project decisions. This is why executives must become involved with projects and interface with project managers to provide project managers with the necessary business information for decision making.

PROJECT ORGANIZATION

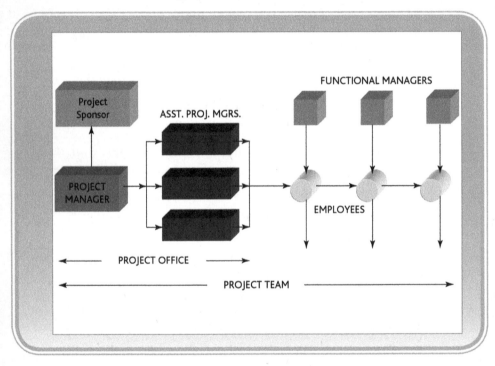

This illustration shows the major players or stakeholders associated with a project.

- The project manager is the person directing the overall project.

- The project manager reports to a project sponsor, who may be at the executive level of the company. The relationship between the project manager and the sponsor is usually a dotted-line relationship.

- On large projects, having assistant or deputy project managers is a common practice. For example, if the project has 10 engineers assigned, then 1 of the 10 engineers may be given the additional title of assistant project manager for engineering. The project manager will now work directly with the assistant project manager for engineering rather than with the 10 engineers.

- The employees are assigned by their respective functional managers and are usually in a dotted-line reporting relationship to the project manager and a solid-line relationship to their functional manager. The selection of an assistant project manager is a joint decision among the project and functional managers.

- For large projects, the project manager and assistant project managers may form and manage a project office.

MULTIPLE BOSS REPORTING

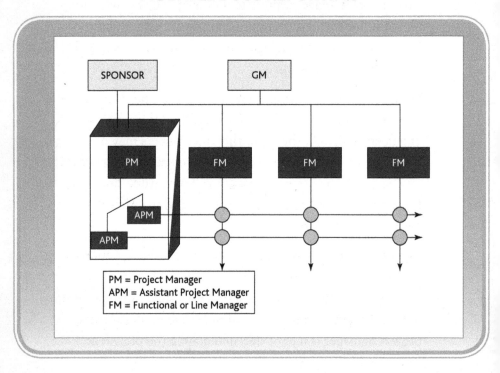

On the surface, it looks as though project management should be easy to perform within a company. Unfortunately, this is not the case. In the illustration on the previous page, the real problem occurs at the circles.

Each circle represents one or more functional employees that must report dotted line to the project manager and solid line to their functional or line manager. This is referred to as multiple boss reporting. The problems occur when the employees receive conflicting instructions from the project manager and functional manager. When this occurs, the employees usually respond to the individual who has the greatest influence on their performance review. This is, in most cases, the functional manager.

Project managers should work closely with functional managers with regard to providing direction to the employees. Placing employees in the middle of a conflict is not a very good idea. Some project managers prefer to provide the instructions to the functional managers first, who in turn will then relay the instructions to the functional employees. While this may incur some sort of small time delay, it does have the benefit of reducing conflicts, as well as keeping the functional managers informed as to what their employees are being asked to do.

PROJECT-DRIVEN VERSUS NON-PROJECT-DRIVEN FIRMS

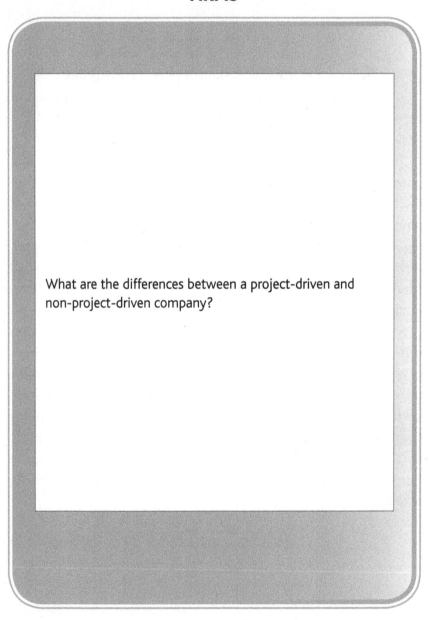

What are the differences between a project-driven and non-project-driven company?

In a project-driven or project-based company, corporate profitability is a result of projects rather than from functional areas. The survival of the company is based entirely on the profitability of projects. In such a case, the functional units exist to support the projects.

In a non-project-driven or non-project-based company, the profitability comes from the work performed in the functional units. The projects exist to support all of the functional units. During a functional crisis, resources may stop working on projects and return to their functional line. Project managers must understand that, in this situation, functional work has a higher priority than project work.

In project-driven companies, project management is regarded as a profession. In non-project-driven companies, it is more difficult to treat project management as a profession because employees associate their future and job security with the functional area rather than through project assignments. Project management may be treated as a part-time occupation in addition to an employee's normal functional responsibility.

COMPLEXITIES IN NON-PROJECT-DRIVEN FIRMS

Why is it more difficult to manage projects in non-project-driven companies?

The difficulties in non-project-driven firms include:

- Projects may be few and far between.

- Not all projects have the same project management requirements and therefore they cannot be managed in a manner consistent with other projects. This difficulty results from poor understanding of project management and a reluctance of companies to invest in proper training.

- An enterprise project management methodology does not exist.

- Executives do not have sufficient time to manage projects themselves, yet refuse to delegate authority.

- Projects tend to be delayed because approvals most often follow the vertical chain of command. As a result, project work stays too long in functional departments.

- Because project staffing is done on a "local" basis, only a portion of the organization understands project management and can observe the system in action.

- There is a heavy dependency on the use of subcontractors and outside agencies for project management expertise.

LEVELS OF REPORTING

Where should the project manager report?

There are both pros and cons to having the project manager report to a high level of management in the organizational hierarchy. The pros include:

- The project manager is charged with achieving results through the coordinated efforts of many functional units. The project manager should, therefore, report to the person who directs all those functions.

- The project manager must have adequate organizational status to do his or her job effectively, which is realized through the relationship with the executive manager.

- To obtain adequate and timely assistance for solving problems that inevitably appear in any important project, the project manager needs direct and specific access to an upper echelon of management.

- The customer, particularly in a competitive environment, will feel a greater sense of support and confidence if the project manager reports to a high organizational echelon.

A major potential problem with having the project manager report to a higher-level or executive manager is that it the relationship may alienate the functional managers on whom the project manager must rely for support.

LOW-LEVEL REPORTING

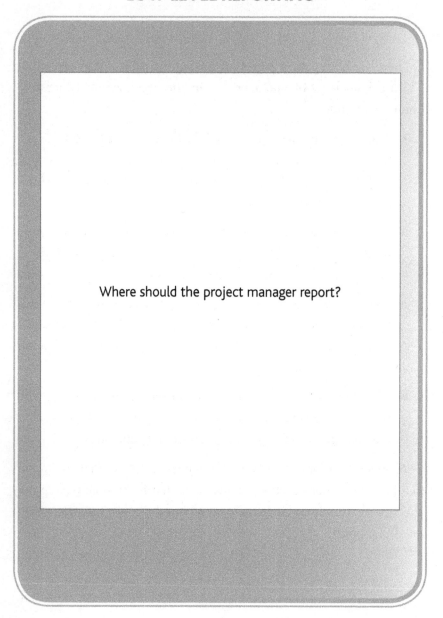

Where should the project manager report?

Having project managers report to a low-level manager in the hierarchy may seem like the right idea, but it does create additional problems:

- It is organizationally and operationally inefficient to be engaged in too many projects, especially small ones, diverting senior executives from more vital concerns.

- Although giving a small project a position of high priority in the organization may create the illusion of executive attention, its actual result is to foster executive neglect of the project.

- Placing a junior project manager at too high a level in the organization may alienate senior functional executives who are relied on to provide support to the project manager.

It is sometimes difficult to find the appropriate position in the hierarchy for a project manager. In project-driven companies, project managers may be positioned at a higher level in the hierarchy than the functional managers, whereas in a non-project-driven company, the project managers are usually at the functional manager level or subordinate to the functional level.

WHY USE PROJECT MANAGEMENT?

WHY USE PROJECT MANAGEMENT?

There are numerous benefits associated with using project management. These benefits affect all functional areas of the company and include:

- Identification of functional responsibilities to ensure that all activities are accounted for, regardless of personnel turnover

- Encouraging continuous improvement and documentation of best practices

- Identification of time limits and risks for scheduling

- Identification of a methodology for trade-off analysis

- Measurement of accomplishment compared to baseline plans

- Early identification of problems to enable appropriate corrective action

- Improved estimating capability for future planning

- Ability to determine when objectives cannot be met or will be exceeded

Later, we will discuss in more detail the benefits achieved from using project management.

WHEN TO USE PROJECT MANAGEMENT

Are there specific situations when project management will work best?

Are there appropriate times when project management appears to work best?

If the answer to any of the following five questions is "yes," then we should consider using project management:

- Are the jobs complex?

- Are there dynamic environmental considerations?

- Are the constraints tight? (extremely limiting)

- Are there several activities and deliverables to be integrated?

- Are there several functional boundaries to be crossed?

The last two questions are often the most critical. When integration is required, project management becomes a necessity and should be utilized. While it is true that functional managers may be able to effectively manage those projects that remain entirely within their functional area, a number of problems may occur when integration requirements cut across multiple functional areas. One functional manager may not have the time or ability to provide the required effort to handle all of the integration issues that may be experienced across an organization while continuing to provide direction to the employees within their functional area.

RELATIONSHIP

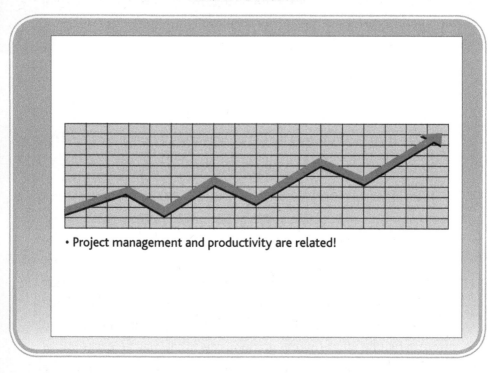

- Project management and productivity are related!

Over the years, companies have discovered that project management and productivity are related. However, the benefits of this relationship do not appear overnight. Some companies require up to five years or longer to develop and install processes and methodologies that will measure true productivity. Productivity may appear in a variety of forms, such as:

- Shortening project time frames

- Lowering of project costs

- Higher levels of quality with no accompanying cost increases

- Fewer meetings

- Fewer conflicts requiring senior management involvement

- Fewer forms, guidelines, checklists, and bureaucratic processes

- Better working relationships between project teams and functional managers

- Repeat business from a multitude of clients

- Higher customer satisfaction ratings

Some benefits can be realized quickly, but the majority of the benefits take time to fully develop.

THE NEED FOR RESTRUCTURING

- Accomplish tasks that could not be effectively handled by the traditional structure
- Accomplish one-time activities with minimum disruption to routine business

If there is a downside risk to project management, it lies in the fact that projects cannot always be accomplished within the existing organizational structure. Organizational restructuring may be necessary, especially in companies that are project-driven. However, regardless of whether the company is project-driven or non-project-driven, there are issues that must be considered:

- Project priorities and competition for talent may interrupt the stability of the organization and interfere with its long-range interests by upsetting the normal business of the functional organization.

- Long-range planning may suffer as the company focuses more on meeting schedules and fulfilling the requirements of shorter-term, tactical projects.

- Shifting people from project to project may disrupt the training of new employees and specialists. This may hinder employee professional growth and development within their fields of specialization.

These issues can be overcome with an effective planning methodology.

IMPROVEMENT OPPORTUNITIES

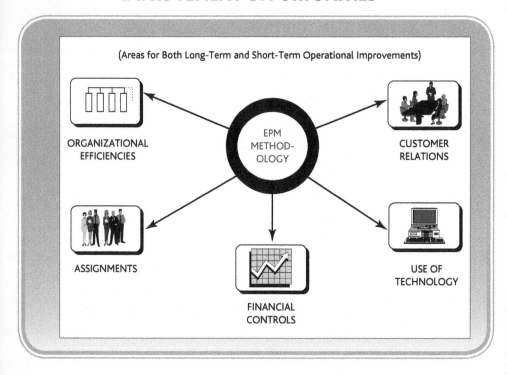

(Areas for Both Long-Term and Short-Term Operational Improvements)

ORGANIZATIONAL EFFICIENCIES

CUSTOMER RELATIONS

EPM METHODOLOGY

ASSIGNMENTS

USE OF TECHNOLOGY

FINANCIAL CONTROLS

The implementation of project management can provide significant opportunities for improvements in various parts of the company:

- *Organizational efficiencies.* Processes can be developed that make organizational work flow more efficiently and more effectively. This can improve profit margins.

- *Customer relations.* Project management allows organizations to work more closely with customers and possibly receive single-source contracts. It also increases our chances for follow-on work from the same client.

- *Assignments.* Assigning people to project teams becomes a more efficient process, and resource capacity planning models can be developed.

- *Financial controls.* Better financial controls will be developed and implemented, both horizontally and vertically. This may necessitate the implementation of an earned value measurement system.

- *Technology.* Technology usage is viewed from a company-wide basis rather from an individual department view.

RESISTANCE TO CHANGE

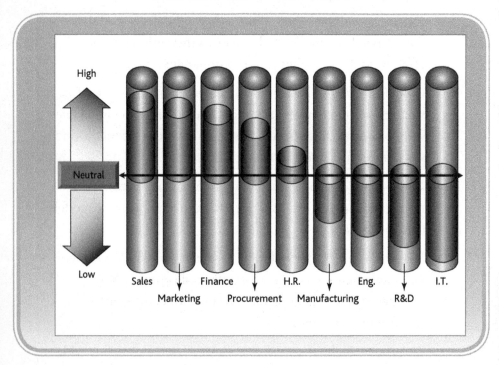

Functional managers and executives must realize that each functional area will form its own view of project management and some resistance to change can be expected. The reasons for resistance might be:

- Sales and marketing view project management as a risk to their traditional receipt of bonuses. They may not support project management for fear of having to share bonuses at successful project completion.

- Financial departments and functional units may feel threatened by the implementation of an earned value measurement system, which may be linked to efficiency reviews. This will require that they learn multiple accounting systems.

- Procurement departments may perceive a decentralization of procurement procedures to the project level.

- Human resources may not see the value of designing a whole new training curriculum for project management.

- Manufacturing, engineering, research and development (R&D), and information technology (IT) may support project management because they see the value and strategic necessity in using it.

Chapter

2

THE BENEFITS
OF PROJECT
MANAGEMENT

BENEFITS OF PROJECT MANAGEMENT

Efficiency:

- Project management allows us to accomplish more work in less time, with fewer resources, and without any sacrifice to quality.

Previously, we discussed some of the benefits of project management implementation. The primary benefit of project management is the streamlining of the work flow which, in turn, allows us to accomplish more work in less time, with fewer resources, and without any sacrifice to quality. This normally requires the implementation of a project management methodology based on user-friendly forms, guidelines, templates, and checklists.

Lessons learned and best practices are captured at the end of each project or at end of phase or phase gate review meetings. The information is then used for continuous improvements to the enterprise project management (EPM) methodology. External benchmarking activities are also effective ways to generate or obtain project management intellectual property for improving the methodology.

When project management is implemented correctly, it works—and it works well. But, as stated previously, there may be some resistance to change introduced by project management. Some people put their own personal aspirations for power and authority ahead of what may be in the best interest of the company. The result can be a very slow maturity process.

Profitability:

- Under normal business conditions, profitability can be expected to increase as a result of using project management processes.

Because project management improves efficiency, profits can generally be expected to rise. For companies that survive on competitive bidding, profitability may very well be based on a series of streamlined processes such as with the implementation of an EPM methodology. Streamlined processes can lead to paperless project management systems that can reduce implementation costs and save companies significant amounts of money.

Executive managers and customers sometimes believe that cost savings will occur if a project manager is not assigned. An example would be a typical information technology (IT) project, where a systems programmer will take on the additional role of the project manager along with functional responsibilities. This situation has a high probability of failure because, without a project manager, there is less focus on the higher-level integrated aspects of the project and less attention to meeting the overall budget and schedule of the project.

Although project management can increase profitability, there are other business decisions that can increase or decrease the expected profits. Many of these other factors are defined as enterprise environmental factors. These factors include organizational culture, tolerance to risk, inflation, recessions, available resources and skill levels, competitive factors, and changes in technology.

Scope Changes:

- Project management advocates effective up-front planning, which should reduce the number of costly scope changes on projects.

For years, customer-funded scope changes were viewed as a source of revenue and added profits. Project managers were pressured to push through scope changes, whether necessary or not, to generate additional income. During competitive bidding, projects were underbid by 20 percent to 30 percent to win the contract with the intent that additional revenue and profit would come from changes that increased scope after the original contract was awarded.

Today, companies are running lean and mean, and do not have the excessive resources that may have been available in the past. While pushing through scope changes may appear to be the right thing to do, resources are required to implement the scope changes. With limited resources, and most of the resources already committed to other projects and ongoing operational work, the acceptance of a scope change may require that resources be temporarily taken from other ongoing projects.

Scope changes have merit if the changes can be shown to add significant value to the customer's deliverables. Pushing through scope changes for the sake of additional profits can alienate the customer and limit follow-on work. Companies should maintain a structured change control process that involves key stakeholders who have the ability to influence change decisions.

Organizational Stability:

■ Project management allows work to flow in a multidirectional manner and usually results in added efficiency, effectiveness, and stability of the organization.

Project management allows work to flow multidirectionally, that is, horizontally and vertically in a concurrent manner. Project managers are generally given the authority to talk to any functional group or business entity within a company about issues concerning the project. Decision making occurs quickly, and less time is wasted going through the often highly bureaucratic chain of command for approvals.

This results in added efficiency, effectiveness, and organizational stability. But this may require that executives allow this to happen by providing the project managers with some degree of authority to make decisions. Organizational stability is achieved when companies make decisions for what is in the best interest of the company before what is in their own personal interest. An organizational culture based on value can make this easier to occur, and senior management is the architect of the culture.

Closeness to the Customers:

▪ Project management allows us to work close to the customers, resulting in a high degree of customer satisfaction.

Project management allows us to work close to our customers, resulting in a greater probability of high levels of customer satisfaction. Customer satisfaction can lead to single-source contracting, which could save hundreds of thousands of dollars by minimizing the need to go through the formalized competitive bidding process.

Some companies are creating EPM methodologies that interface directly with the customer's project management methodology. After project closure, the customers are interviewed to see what changes they might recommend to the existing methodology for use on future projects contracted with the customer. Since customers are now tracking their project through the contractor's project management methodology, contractors are providing training for customers about how the methodology works.

Problem Solving:

■ Project management allows for better problem solving and usually in a shorter period of time.

Project management can provide a basis for better, more informed decisions made in shorter periods of time. But this requires senior management to provide project managers with sufficient authority (based on type of project and importance to the company) for decision making. This may not occur until the organization achieves some degree of project management maturity.

The concept of integrated project teams (IPTs) has become popular during the last decade. IPTs are based on (1) selecting the appropriate team members that possess the technical know-how and (2) providing the team with the authority to make decisions. This concept of using IPT has been shown to accelerate decision making and reduce the project's life-cycle cost and completion time. Companies that have used this technique beneficially include Hewlett-Packard, 3M, and DaimlerChrysler.

When the team possesses the knowledge and authority to make decisions, the team requires significantly less interfacing with organizations external to the project. Less time is spent interfacing with sponsors for decision making and seeking out subject matter experts that must ramp up to become familiar with the technical issues.

Application:

- Project management processes can be applied to all projects, regardless of the size and scope of the projects.

Historically, project management was used only for implementing large, complex projects or those above a certain threshold limit such as:

- Exceeding a certain dollar value

- Exceeding a certain time duration

- Requiring integration across a certain number of functional units

- Special customer requirements

- Project risks

Today, we believe that companies should develop an EPM methodology. This methodology should be used on all projects regardless of the size and scope, but only those portions of the methodology that are appropriate for the type of project should be used. Generally, EPM methodologies are adjusted to meet specific project needs. It may not be cost-effective for the entire methodology to be used on all projects.

Establishing threshold limits is usually a bad idea because someone in a position of authority can arbitrarily change the threshold limits and the project management methodology will no longer be effective and will not be used. Methodologies allow problems to surface quickly rather than being buried, and also create early warning indicators that problems are about to surface.

Quality:

- Effective use of project management processes can increase quality without an accompanying increase in cost.

Effective project management can improve quality without any accompanying increase in cost. Most people do not fully understand the connection between project management and quality. People seem to understand that project planning may very well be the most important life-cycle phase of a project, at least through the eyes of the project manager. But what is the most important life-cycle phase when considering quality?

Quality is *not* inspection. Quality must be planned for and, therefore, planning could be regarded as the most important phase for quality. The connection and continued relationship between project management, quality, and possibly Six Sigma occurs during the planning process.

The EPM methodology is a structured process. Using the methodology provides some degree of structure to the implementation of quality. Six Sigma project managers must recognize the importance of using a statement of work, work breakdown structures, and schedules.

Authority Issues:

- Contrary to popular belief, project management actually reduces authority and power issues in companies.

Effective project management is actually leadership with limited or, in some cases, no authority. In virtually all companies, the real authority is already carved up and shared among the executives, middle management, and some lower-level managers. It is unrealistic for someone to believe that, simply because he has been assigned as a project manager, he will be given a charter for the project containing a vast amount of authority delegated to him by senior management.

Most project managers seem to have more implied authority than real authority. The real authority resides with senior management and the project sponsors. Once this is recognized, people learn that project management is actually leadership without authority, and conflicts over power and authority are diminished.

However, many conflicts can be resolved if the authority granted to the project manager is documented. Employees must know what level of authority the project manager possesses with regard to decision making. Likewise, project managers must know how much authority the project team members have with regard to making decisions for their respective departments.

Cost of Using Project Management:

- The cost of implementing and using project management is low, and the result should be an increase in business rather than a decrease.

Four decades ago, companies refused to consider the use of project management for fear that the cost of implementation would make the company noncompetitive. There was a very real fear that new layers of management would be needed to provide supervision for project management activities. This would be costly and significantly drive up the overhead costs.

Today, many organizations have realized that the supervision costs are not necessary, and the project sponsor can fulfill the supervision and support role. However, there may be an initial cost increase as project management processes are implemented and an EPM methodology is created. There may also be an initial cost for project management training and education as well. Within a year or two after implementation, the organization should become more efficient and more effective in the way it executes project work, and the costs associated with using a project management methodology should be reduced considerably.

There is no doubt that a return on the investment in project management will be realized when introduced and applied. However, the return may not be readily apparent until two to five years after implementation. Like most activities, there are learning curve effects to be considered.

Solution Provider:

■ Project management processes allow companies to function as solution providers rather than merely as suppliers of products or services.

When companies become efficient or excel in the use of project management methodologies, they capitalize on this expertise by marketing themselves as business solution providers rather than simply as the seller of products or services. As solution providers, companies sell:

- Their project management skills

- Their ability to interface operations with the company's project management methodology

- The ability to share knowledge documented in their best practices library

In exchange for providing customers with business solutions, the company's goal is to develop a relationship in which their customers treat them as a strategic partner rather than just a contractor or supplier of goods and services. This concept of being a solution provider is now part of engagement sales and marketing activities. Solution providers focus on the long term rather than the short term; follow-on work rather than single sales; selling long-term value rather than specific features; selling quality of the solution rather than quality of the products; and viewing customer service as a high priority rather than a low priority.

Suboptimization Risks:

- Project management allows us to make decisions that are in the best interest of the company as well as the project.

Some individuals in key company positions may have the tendency to make self-serving decisions, focusing more on their own best interests than what is in the best interest of the company. Project managers and project teams must make decisions not to satisfy their own needs but in the best interest of achieving the project objectives, the stakeholders, and the company goals. However, this is not easily done. To ensure effective decision making, team members must understand the business needs as well as the technology used to manage projects. Depending on the level of business knowledge contained within the team, there may be a greater need for close interfacing between the executives, the project sponsor, and the project team.

It is the responsibility of the executive project sponsor to make sure that the team understands the business implications of the project. The sponsor also defines the business objective for the project, and the sponsor must visibly support the project manager's efforts to ensure that the team makes appropriate decisions.

3

SOME IMPLEMEN- TATION COMPLEXITIES

THE CHALLENGES FACING PROJECT MANAGERS

- Project complexity
- Customer's special requirements and scope changes
- Organizational restructuring
- Project risks
- Changes in technology
- Forward planning and pricing

Previously, we stated that one of the major challenges facing project managers occurs when resources that report directly to the functional manager may be assigned to several projects. But there are other complexities that also impact the functional managers:

- Projects are considered to be unique activities. As such, the complexities of the project may make integration of work across functional interfaces very difficult, requiring significant help from functional managers.

- Sometimes projects are accepted with special customer requirements. These requirements may be totally new to the functional managers, and additional time may be required to perform the work. Lengthening of schedules may be necessary to accommodate the demands placed on the functional managers.

- Functional manager estimates for cost and time are often based on the functional manager's experience and tolerance for risk. Some functional managers have a high utility for risk (they are risk seekers and thrive on risk taking) and provide optimistic estimates, while others are risk averters and provide extremely pessimistic estimates.

- Changes in technology or new technologies may add risks to functional manager responsibilities, especially if the functional managers are unfamiliar with the technology.

- Long-term projects, generally three years or longer in duration, require that the functional managers estimate the salary of the workers, availability of the workers, and skill levels of the workers well into the future. This is also further complicated if the functional manager leads a functional unit where technology changes rapidly.

WORKING WITH THE TECHNICAL PRIMA DONNA

Project managers are at the mercy of the functional managers when it comes to the quality and skill level of the assigned functional employees. Project managers can request specific individuals with desired skills, but the final decision about resources rests with the functional manager.

Some functional employees have outstanding technical strengths but simply do not work well in teams. These people work well by themselves and often resent supervision. It is the responsibility of the functional managers to inform the project managers about these people during the project staffing process.

Characteristics of a technical prima donna (a vain and temperamental person, a disagreeable person, an unpleasant person) include:

- A desire to work alone

- A desire to work without close supervision

- When placed in charge of a team of people, will do all the work himself or herself and have little faith in the results of the team

- Must always validate other people's solutions before accepting them

Functional managers must provide some buffering to make it easier for the project managers to work with this type of person given the fact that the project manager may have no choice in the selection of resources.

EARLY REASONS FOR FAILURE

Although rare, why does project management sometimes fail?

It has been proven on countless projects that project management works, and works well. But in some situations, project management methods falter and may even fail. Such situations include:

- There is no perceived need or desire for project management.

- Employees were not informed about how project management should work or trained in the basic principles.

- Functional managers were not informed about how project management should work.

- Functional managers did not understand their role in project management.

- Executives did not select the appropriate projects or project managers for the first few projects.

- There was no attempt to explain the effect of the project management organizational structure on the wage and salary administration program.

- Employees were not convinced that executives supported the change to project management.

- Management selected the wrong type of project to serve as a breakthrough project to convince people that project management would work.

- There was no attempt to clearly articulate the benefits that would be achieved as a result of using project management.

Chapter

4

ROLE OF THE MAJOR PLAYERS IN PROJECT MANAGEMENT: THE PROJECT MANAGER

THE THREE-LEGGED STOOL

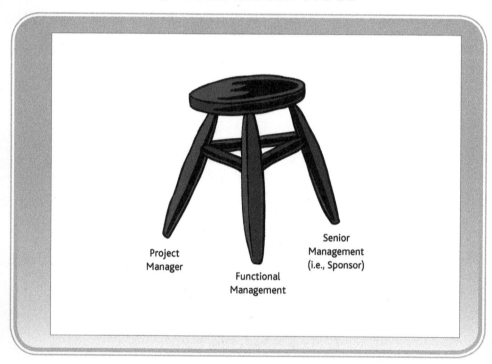

Previously, we discussed some of the key players that are associated with project management. Here, we are viewing project management as though it functions as a three-legged stool. The three legs are:

- The project manager
- The functional managers
- Senior management

To ensure a solid, balanced approach and that project management functions as planned and achieves the expected benefits, all three legs of the stool must understand project management and how it should function. If only two of the legs understand project management, it is impossible for the stool to stand.

In the following pages, all three of these legs will be discussed.

THE PROJECT MANAGER'S STOOL

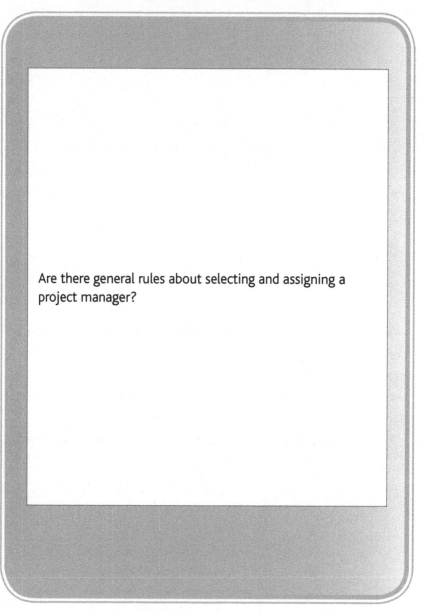

Are there general rules about selecting and assigning a project manager?

The project manager is the person who is ultimately held accountable for the successful completion of the project. Simply stated, the project manager must coordinate the activities necessary to develop, execute, and apply necessary changes to the project plan to achieve success. To accomplish this, the project manager must work with a variety of employees and stakeholders, all at different levels in the organizational hierarchy.

With this in mind, we can identify a reasonable criterion for the selection of the project manager:

- A project manager is given the authority to manage across several organizational lines. His or her activities, therefore, resemble many aspects of general management, and must be performed well to continue to receive managerial support.

- Project management will not succeed without effective project managers. Thus, if executive management approves a project for implementation, it should certainly ensure that a qualified and reliable person is selected as the project's leader.

- A project manager is far more likely to gain the support of the project team and functional managers to accomplish desired goals, if it is obvious that executive management has selected, appointed, and visibly supports the project manager.

NEGOTIATING FOR RESOURCES

Project managers are responsible for clearly identifying the project requirements, which will provide the functional managers with the information needed to determine the appropriate resources and skill levels necessary to accomplish the deliverables. Project managers may request specific people, based on working relationships established during previous projects, but the functional managers may not necessarily support or agree to the request.

Even in cases where the project manager may have a command of technology, it is usually the functional manager who makes the final decision regarding resource assignments. Functional managers generally have a better knowledge of the skill levels and capabilities of the workers than the project manager.

Project managers should avoid involving the project sponsor during the resource negotiation process. Sponsors are unlikely to usurp the authority of the functional managers in the resource assignment process. Sponsors and executives would normally become involved when resource negotiations between project manager and functional managers breaks down or reaches an impasse. Asking sponsors to intervene too early can create conflicts between the project manager and functional manager that are not easy to resolve.

THE PROJECT KICKOFF MEETING

Are there general rules about selecting and assigning a project manager?

The typical launch of a project begins with a kickoff meeting led by the project manager and includes the major players or stakeholders who will be responsible for planning and executing the project. These stakeholders include the project manager, assistant project managers for certain areas of knowledge, technical subject matter experts (SMEs), and functional leads.

There may be multiple kickoff meetings based on the size, complexity, and time requirements for the project. The major players are usually authorized by their functional areas to make decisions concerning timing, costs, and resource requirements.

Some of the items discussed in the initial kickoff meeting include:

- Wage and salary administration, if applicable

- Guidelines for performance appraisals

- Initial discussion of the scope of the project, including both the technical objective and the business objective

- The definition of project success

- The assumptions and constraints as identified in the project charter

- The project's organizational chart (if known at that time)

- The participants' roles and responsibilities

- Known project risks and contractual agreements

ORGANIZING THE PROJECT TEAM

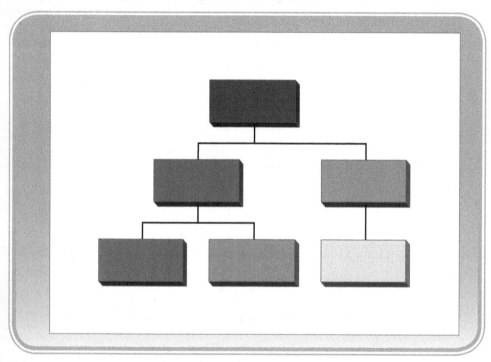

The project manager, working with the functional managers and possibly the project sponsor, determines the size and leadership requirements for the project team. Some of the factors influencing the design of the project team include:

- Size and complexity of the project

- Funding for the project team

- Technical requirements of the project

- Relationship to business objectives

- Visibility of the project internally and externally

- Political sensitivities

- Size of the customer's project office

Large projects are often managed by creating a project office that is staffed with assistant or deputy project managers and support personnel. For example, if the project requires a large number of engineers to be assigned, one of the engineers may be assigned as the assistant project manager for engineering. This arrangement will help to control the project more effectively and assist with integration of project work. The functional manager must agree that the selected engineer can effectively manage the additional responsibilities.

When working with external customers, it may be necessary during the implementation of large projects for the contractor's project office and the customer's project office to be of comparable size and responsibilities. Some customers need to be "catered" to on a regular basis, and, to use a sports metaphor, "man-to-man" coverage often works better than a zone defense.

RESPONSIBILITY ASSIGNMENT MATRIX

LEGEND

○	General management responsibility
●	Specialized responsibility
◀	Must be consulted
◀	May be consulted
▨	Must be notified
□	Must approve

Raw Material Procurement

Task	Project Sponsor	Department Manager	Team Member	Project Office	Project Manager
Prepare bill of materials		◀	○	◀	
Contact vendors		□	○	◀	
Visit vendors		□	●	○	○
Prepare purchase orders	□		◀	○	
Authorize expenditures			□	□	
Place purchase orders			◀		
Inspect raw materials		○	◀	□	
Quality control testing		○	◀		
Update inventory file		○	◀	□	
Prepare inventory report		○	◀	◀	
Withdraw materials		○	○	□	

The illustration on the previous page displays a responsibility assignment matrix (RAM), also called a linear responsibility chart or a RACI chart (Responsible, Accountable, Consulted, Informed). The intent of the RAM is to make sure that each project team member clearly understands his or her role and responsibility during the project assignment. Because employees may be assigned to several projects at the same time, it is essential that the employees have a clear understanding of what they are expected to do for each project, including activity assignments, information sharing, authorizing decisions, and approvals.

Project managers usually complete the RAM, which should include the input and approval of the functional managers. Functional managers may not want their employees to be assigned to certain levels of responsibility above their current pay grade. There may be a concern that the responsibility assignment may be viewed by the employee as an indication of possible promotion.

The RAM is often prepared using the names of the employees rather than their job titles or position. This may lead to problems if the customers have the names of all of the team members and decide to contact these people directly without following protocol or going through the project manager or project office first. For projects intended for customers external to the firm, the RAM may be prepared with the titles of the employees rather than the names. There are certain situations and contractual terms that require specific individuals by name. Care should be taken when negotiating situations of this nature.

ESTABLISHING THE PROJECT'S POLICIES AND PROCEDURES

Projects often have unique characteristics that are included in the company's standard policies and procedures. As a result, project managers may find it necessary to develop policies and procedures unique to the project. These policies and procedures must not violate company or functional policies and procedures or collective bargaining agreements. Therefore, project sponsors and functional managers may find it necessary to review the project's policies and procedures to ensure that they are acceptable and do not violate other established organizational policies.

Sometimes, rather than develop new policies and procedures, project managers may focus on interpreting company policies and procedures in their own best interest. The role of the sponsor is to assist with the interpretation of policies to ensure that project interests are addressed while maintaining organizational integrity. The sponsor also serves as a "safety net" for the project manager should any of the policies require a specific interpretation in connection with a benefit that may be associated with the project. The project managers may also have to adhere to some of the client's policies and procedures, and the sponsor can assist in this interpretation as well. Functional managers may also become involved in the development and/or interpretation of policies as well.

Today, most companies are in or working toward a paperwork reduction mode and focus heavily on the development of forms, guidelines, templates, and checklists that will improve efficiency and reduce bureaucracy. These documents may also require interpretation, and the sponsor and functional managers can be of assistance here as well.

LAYING OUT THE PROJECT WORK FLOW AND PLAN

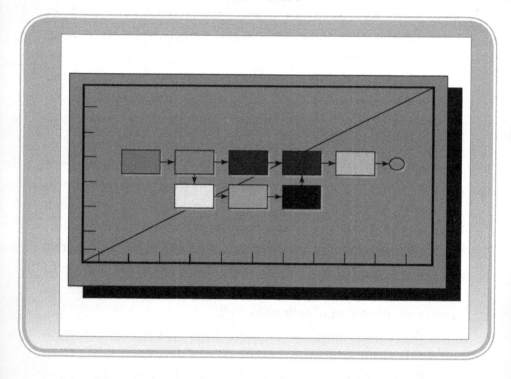

Once the project management plan—namely, the organization of the project team and how they will work together—is developed, the next step is the preparation of the project's master plan. This is a high-level plan including critical milestones established internally and by the customer.

The sponsor's involvement in the daily activities of the project is on an as-needed basis as issues arise. The sponsor also validates the high-level plan to make sure that all critical milestones are identified. The master plan is usually a milestone chart but can also appear in the form of a high-level network diagram with time-phased milestones superimposed on it. The master plan identifies the critical milestones that will be defined in the detailed planning process.

The functional managers should agree to the milestones in the master plan. If the functional managers cannot accomplish the work according to the project manager's deliverables, the milestones may require adjustment or changes. This could be an iterative process involving the project manager, project sponsor, customer, and functional managers. In some cases, contractual agreements limit the ability to change milestones.

The master plan may require the approval of the customer to make sure all critical milestones have been identified. The presentation of the master plan to the customer may be made by the sponsor as part of the executive-to-executive communications or by the project manager, but with functional support to address technical issues and questions.

ESTABLISHING PERFORMANCE TARGETS

The major objectives for the project are established prior to project initiation and during meetings between the customer, the project sponsor, and, if assigned, the project manager. However, there may be hidden or secondary objectives established above and beyond the customer's objectives. As an example, a company wins a contract to manufacture some products for a customer. On the surface, it may seem that the objective is simply the manufacturing of high-quality products for a customer. But now let's assume that, in order to win the contract, the company bid and won the contract at a price 20 percent below its own cost of performing the work. The secondary objectives, not identified to the customer, might be to keep its people employed, establish a foothold for future contracts from this customer, develop new manufacturing processes, or maintain a specific market share.

Functional managers must be informed of all project objectives. Secondary objectives may force functional managers to accept risks and complete the project in less time than normally expected. This places exceptional pressure on functional managers.

Not all secondary objectives are shared with the project team. There may be a "need to know" type of situation where the managers do not share all information due to sensitivity of the information or its impact on project or operational work. Telling the team that the secondary objective was simply to keep people employed may create fear that layoffs are possible in the future. Sometimes, secondary objectives are shared just between the project sponsor and the project manager.

OBTAINING FUNDING

One of the first responsibilities of the project manager is to obtain funding for the project. For projects that are external to the company, funding is provided by the external customer. If the project manager is forced (usually by the sales team) to accept a budget that the project manager deems to be inadequate, then pressure is usually placed on the functional managers to perform the required work within the constraints of the budget. If this happens continuously, work quality may be affected and the functional managers may begin adding significant padding to their estimates.

For projects internal to the company, funding is usually provided by the sponsoring organization. The target budget may be established well before the project manager is assigned. Once the plan is prepared and cost estimates have been developed, the project manager must determine whether the preestablished budget is sufficient to meet the requirements. It is generally easier for the project manager to obtain an increase in the budget for internally funded projects.

Some project managers prefer to have a management reserve as part of the funding. The management reserve is a sum of money added into the budget to compensate for escalations in salaries, overhead rates, and procurement costs over the life of the project. The management reserve is not to be used for scope changes. The magnitude of the management reserve is based on the project's overall risk. Typical management reserves range from 2 percent to 15 percent. However, some people use contingency reserve as the sum of money included in the budget because escalations may be regarded as known unknowns. In this context, management reserve is for unknown unknowns.

EXECUTING THE PLAN

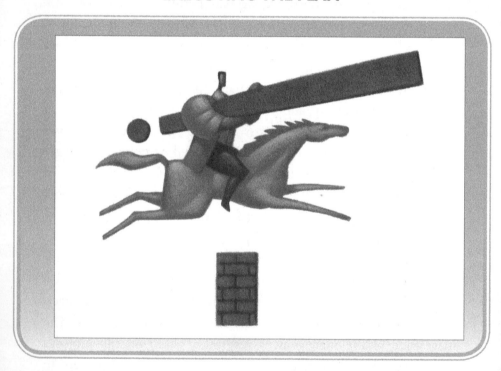

Project managers, working closely with functional managers, are responsible for the execution of the project plan. Project performance is reviewed routinely by the project sponsor to make sure that work is progressing as planned. Sponsors generally have higher-level management responsibilities that require their attention and consume most of their time. Therefore, sponsors do not have sufficient time to become actively involved in the details of the project, especially after project execution begins.

Sponsors are expected to monitor project progress from a high level of the work breakdown structure (WBS). Project managers, and possibly functional managers, meet with the sponsor on a regular, often weekly basis to provide high-level briefings. While it may seem appropriate for sponsors to attend team meetings, this may not be an effective use of the sponsor's time, especially if the executive is sponsoring multiple projects concurrently.

Project managers do not usually have authority over the assigned functional resources. Project managers must rely on the functional managers to provide the deliverables to which they committed during project planning. If the team cannot meet a project deliverables, the project and functional manager must jointly explain to the sponsor and other executives why the deliverable could not be met and offer solutions to the issues preventing the completion of the deliverable.

ACTING AS THE CONDUCTOR

From the day the project manager is assigned, he or she acts as the conductor of the project team. The role of the project manager is to make sure that the work is proceeding in harmony. But, unlike a conductor who has a command of the technology and components involved, the project manager may not possess a detailed and thorough understanding of the technology. When this occurs, the project manager must rely on the functional managers for conducting that portion of the work that resides within the functional manager's area of expertise.

PUTTING OUT FIRES

Project managers are expected to make the necessary decisions to manage crises. Sometimes, this is difficult to do without having a complete knowledge and command of technology. In such cases, project managers may find it necessary to rely on the functional managers and the project sponsor for assistance.

Decision making in most companies is a tedious process. Because project managers may have limited authority, they may be at the mercy of others when it comes to decision making. Also, project managers may not be familiar with decision-making techniques and criteria, especially those in the functional areas.

Employees and executives alike may not realize the importance of the time constraint and its impact on a project. Project managers realize that time is a constraint, and responding to and resolving "fires" may have to happen quickly. The project sponsor has the authority to accelerate the problem-resolving process to minimize the impact on the project. On one government contract, a "Unit Manning Document" existed stating the number of functional employees that could be assigned to the contract and the expected skill levels of the employees. The project manager was in desperate need to have an additional person assigned to the project. It would have taken the project manager six months to have the Unit Manning Document changed. The sponsor worked directly with the human resources executive and accomplished the change in two weeks. The conclusion is that the project manager must know when to ask for assistance and also know his or her limitations with regard to putting out fires.

COUNSELING AND FACILITATION

Problems can and will occur during project implementation. Sometimes, employees make mistakes or simply do not understand the job well enough and then seek out assistance from the project manager. Employees often prefer counseling from the project manager rather than the functional manager because it is less likely that their mistakes would be documented officially and used during performance reviews.

Project managers must understand human behavior in order to perform counseling and facilitation correctly. Facilitation techniques are essential skills for project managers because team meetings can lead to heated discussions and conflicts. Therefore, project managers must develop and maintain some degree of effective communication and interpersonal skills to perform their job effectively.

ENCOURAGING THE TEAM TO
FOCUS ON DEADLINES

All projects are subjected to the triple constraint that emphasize deadlines, baselines, and the deliverables that must be accomplished. Assuming that the deadlines are realistic, the project manager, project sponsor, and functional managers must continuously reaffirm to the project team the importance of meeting deadlines. Unfortunately, this is often difficult to accomplish because of changing priorities, the loss of critical resources, risk events, and other unforeseen problems.

One possible solution is the use of the principles of risk management and what-if scenarios, and through the development of contingency plans. Team members must be encouraged to identify and communicate problems quickly, especially issues such as unrealistic deadlines, to ensure that the maximum number of options can be generated for use in the development of contingency plans. Some companies encourage problems to be brought to the surface as quickly as possible so that corporate support can be made available as quickly as possible. People should be encouraged to report bad news and problems and must not be punished or reprimanded for bringing forth bad news.

Deadline commitments require periodic reevaluation. Trade-off decisions may be necessary to meet the initially agreed-upon deadline.

MONITORING PROGRESS BY "POUNDING THE PAVEMENT"

Reports, no matter how detailed, do not always show the true status of the project. Real status can be seen using the walk-the-halls or management-by-walking-around concept. In addition to seeing the true status, another benefit is the motivational impact this approach will have on team members when they see the interest level of the manager. This approach is particularly effective when sponsors and executives are involved.

Team members want to believe that the project manager and functional managers are sincerely interested in the work they do. Not all team members attend project team meetings and, as a result, may work on a project for an extended period of time and never see the project manager. Some type of communication with these team members is extremely important to maintain morale and sustain project loyalty and commitment.

EVALUATING PERFORMANCE

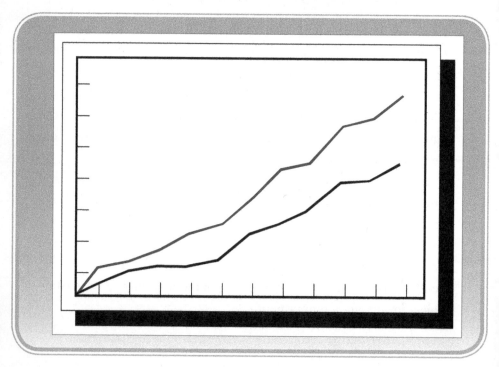

After project initiation and planning is completed, the role of the project manager becomes more associated with monitoring and (some) control and, of course, conflict resolution and problem solving. As part of monitoring and control, the project manager must measure progress and prepare the appropriate reports. The project manager must:

- Measure resources consumed

- Measure progress and accomplishments

- Develop progress and status reports

- Compare measurements to projections and baselines

- Determine the variances from the baselines

- Determine the cause of the variances

- Work with the functional managers to develop contingency plans or actions to reduce the unfavorable variances

- Implement corrective actions and new or revised plans

- Develop project forecasts

- Repeat the measurement process throughout the project life cycle

In most cases, the information for project reports must be collected from a variety of functional managers, and their input must be reliable and accurate.

DEVELOPING CONTINGENCY PLANS

Today, more and more companies are using portfolio management techniques for selecting, grouping, and determining the sequence in which projects will be implemented and how resources will be used. Many times, the portfolio selection process is based on capacity planning activities, where the projects selected are matched against the availability of specialized subject matter experts.

Project managers may have limited knowledge about capacity planning activities, portfolios, and the project selection process, and functional managers may not be involved in the selection process. What is unfortunate about this situation is that projects may be selected without input from the project manager, which may lead to planning and execution issues and potentially more risk. Another issue to consider is that the selection process may have already predetermined what resources will be assigned to a given project. If the assignment includes a critical resource (specialized skill but limited availability or number) that is unavailable for assignment or if the resource is assigned but later removed and reassigned to a higher-priority project, the project manager must develop contingency plans with functional managers to address the resource issue.

Generally speaking, there is more flexibility in the development of contingency plans related to changes in scope, quality, and budget. The process of developing contingency plans due to the loss of critical resources is very difficult and must include participation of the functional managers and project sponsors.

BRIEFING THE PROJECT SPONSOR

Project managers must brief the sponsors about the project condition on a regular basis. Sponsors expect to receive high-level, summary-type project status reports on a regular basis. Failure to provide the expected status or inconsistencies in the reports may result in micromanagement from the executive levels.

For high-level sponsor briefings, two main questions must be addressed:

- What is the condition of the project today with regard to time and cost?

- Where will we end up (forecasts) with regard to time and cost?

Another reason for briefing the sponsor is the potential for media attention. Based on the type of industry or nature of the project, the larger the project, the greater the attention provided by the news media. It has been said that bad news sells more newspapers than good news. Schedule delays and cost overruns on highly visible projects are very likely to make the front page.

Project sponsors are generally positioned much better than project managers and may have more suited skills and authority to address the media in time of crisis. Project managers view crises from a project perspective, whereas sponsors look at the same issues from a company or business perspective. Also, many customers expect to hear about critical problems first from the sponsor rather than the project manager. The reason for this is that the customer feels some degree of satisfaction and confidence when senior management is involved and on top of the issue.

REVIEWING STATUS WITH THE TEAM

One of the primary purposes of project team meetings is to report project status, determine the condition of the project, and share the information with the stakeholders. People who work on project teams become highly motivated when they see how their efforts and contributions fit into the big picture. Therefore, providing all stakeholders with a composite picture of the status of the project is highly recommended.

Project team meetings may include a discussion about every open work package in the work breakdown structure (WBS). Some work packages may be discussed during one-on-one sessions or small group discussions between the project manager and functional manager or with the employee(s) involved if the project manager feels that it would be more productive than to have an entire team sit through the discussion.

Another purpose of team briefings is to discuss future or potential risks and to assign required action items for review at the next team meeting. Action items may be related to new activities that the team members had not anticipated or may be necessary to resolve an unforeseen issue. Action items may not be directly attributed to specific deliverables but may be necessary to keep the project moving forward smoothly.

BRIEFING THE CUSTOMER

Project managers do not necessarily brief each and every executive in their own company on the status of the project. In most cases, the project manager briefs the project sponsor, who in turn briefs other executives. One of the reasons for doing it this way is to provide summary information to executives. Project managers tend to provide project details that may not be of interest to executive management. The project sponsor provides a "detail buffer" between the project manager and the executive team and communicates the essential information the executives should be aware of.

Customers, however, are interested in the details and often demand to have all project information available for discussion and analysis during status or briefing sessions. What may not be apparent to the project manager is that problems with the project being reviewed may actually have a serious impact on the organization's budget or other projects under way in the customer's company, which justifies the interest in the details.

Because of the customer's interest in the details of the project, customer interface meetings are frequently held at the contractor's company location to arrange for functional managers and their subject matter experts to attend or to be easily reached if necessary. For large high-technology projects, a customer interface meeting may be held in a formal conference setting, with functional managers presenting information directly to the customer instead of presentations by the project manager or the assigned team members.

CLOSING OUT THE PROJECT

Once the customer agrees that all deliverables have been provided and accepts the project as complete and successful, the project will proceed to the closing phase, often referred to as contractual closure. The project manager must begin the process of administrative and financial closure. Financial closure includes the closing of all cost accounts, charge lines, and all open work orders. Administrative closure includes:

- Completing and archiving of project reports, memos, newsletters, and project correspondence

- Making sure that priority or confidential information is safeguarded

- Documenting all scope changes to ensure that an audit trail exists

- Debriefing the project team to capture lessons learned and best practices

- Asking the team members to update their resumes in the company's skills inventory data base

- Releasing team members back to their functional departments

- Preparing for financial payments from the customer

Planning for the closure of the project should be accomplished with the same enthusiasm that was demonstrated by the project and team during project initiation and planning. In some companies, the project manager is replaced with another project manager whose main expertise is project closure.

PROJECT MANAGEMENT SKILLS

Given the activities that the project manager must perform, what skills should the project manager possess?

The skills needed to perform as a project manager may be different from company to company and project to project. However, some common skills and characteristics among project managers might include:

- Demonstrating honesty and integrity with all stakeholders

- Understanding of the business

- Understanding of project technology, but not necessarily a command of technology

- Coping skills—ability to handle stress and pressure

- Understanding of team dynamics

- Effective communication skills

- Organizational skills and discipline

- Ability to identify problems

- Willingness to make decisions

- Demonstrating a tolerance for risk

- Delegation

- Negotiation

The relative importance of these skills can change based on the size of the project, the nature of the project, the quality of the resources that are assigned, the importance to the customer, the importance to the contractor, and even the project manager's personal ambitions.

What are some of the skills that may be unique to a given project?

Some projects may require that the project manager possess specialized skills. Some of these specialized skills include:

- Determining if a feasibility and economic analysis are necessary

- Determining the need for complex technical expertise

- Determining if there is sufficient backup strength in the functional organizations and assessing organizational capability

- Determining the risks associated with new and complex technology

- Determining the priority of the project compared to other projects and business needs

- Assessing interface requirements internally and externally

- Ability to manage several contractors at the same time

- Ability to deal with the media

- Ability to work with various government agencies

- Qualifications to obtain the appropriate security clearance that may be required by the company, government, or special agencies

5

ROLE OF THE MAJOR PLAYERS IN PROJECT MANAGEMENT: THE PROJECT SPONSOR

THE NEED FOR A SPONSOR

Most projects also have a project sponsor that may or may not reside at the executive levels of management.

Because project managers and functional managers do not always agree, and because unknown problems occur that cannot be resolved by the project manager, a project (or executive) sponsor is may be needed. The role of the sponsor can change, based on the specific life-cycle phase of the project. For example, during the project initiation and planning phases, the sponsor may take on a very active role to ensure that the proper objectives are established and that the project plan satisfies the needs of the business as well as the needs of a particular client. During the execution phase, the sponsor may take on a more passive role and become involved on an as-needed basis, such as when roadblocks appear, crises develop, and conflicts exist over resources and priorities among projects.

There are other factors that can determine the involvement of the sponsor. These include:

- The risks of the project

- The size, scope, and nature of the project

- Magnitude of the conflicts and obstacles

- Who the customer is

- Specialized customer requests or exceptions

- Quality and availability of assigned resources

- Priorities among projects

- Requests for scope baseline or schedule baseline changes

THE PROJECT SPONSOR INTERFACE

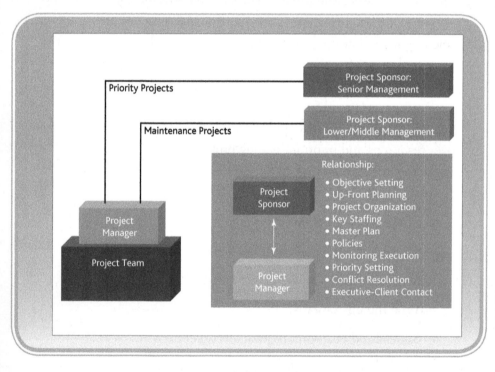

All projects should have a sponsor assigned, but not all projects require that sponsorship occur at the executive levels. The illustration on the previous page shows the relationship between the project sponsor and the project manager, and also shows that:

- Sponsorship is not always positioned at the executive level.

- The sponsor supports the entire project team, not just the project manager.

- Many of the activities in the "Relationship" column can be a shared responsibility between the project manager and the sponsor. However, the decision as to who actually would be responsible for an activity can vary, depending on the life-cycle phase.

The role of the sponsor, in addition to providing funding and clearing away roadblocks that may appear during project execution, is to make sure that the business interests of the company are addressed. If an official sponsor is not appointed, the project manager's immediate supervisor may act as the sponsor by default. The sponsor also acts as a facilitator between project and functional managers to ensure a collaborative work environment.

6

ROLE OF THE MAJOR PLAYERS IN PROJECT MANAGEMENT: THE FUNCTIONAL MANAGER

CLASSICAL MANAGEMENT

- Planning
- Organizing
- Staffing
- Controlling
- Directing

Which of the above is usually *not* performed by the project manager?

A t this point, we will begin the discussion of the role of the functional manager. On the previous page, we identify the five primary functions of classical or traditional management. On the surface, it appears that both the functional manager and project manager perform all of these functions. In reality, the functional manager does all of these, but the project manager performs only four of these. The project manager generally does not staff the project. The project manager negotiates with the functional manager to arrange for staffing.

Project managers have the right to request specific people, especially if the project manager has a good working relationship with these people or if a specific technical skill is required. However, the functional manager makes the final decision. Project managers must realize that the functional managers have the responsibility for the staffing of several projects and the requested resources may be needed on other, higher-priority projects.

Some people argue that project managers should simply identify the grade level or skill set of the resources needed. However, to do this would require the project manager to possess a very thorough knowledge and command of technology to assess resource needs.

THE FUNCTIONAL MANAGER'S ROLE

Exactly what is the functional manager's role during staffing, and why do conflicts with the project manager exist?

In cases where the required project deliverables are not met, the inevitable blame is placed on either the project manager or the functional manager. The project managers may argue in their defense that the wrong resources were assigned by the functional managers. The functional managers argue that the resources were used incorrectly or the wrong resources were assigned because the project managers did not provide a clear description of the project's resource requirements. We can clarify the role of each as follows:

- The primary role of the project manager is to coordinate and integrate project activities.

- The functional manager has the responsibility to define *how* the project tasks will be done, who will perform the work, and the technical criteria associated with the task.

- The functional manager has the responsibility to provide sufficient and qualified resources to perform project activities and accomplish the objectives within the project's constraints.

Effective up-front planning and a good scope statement can minimize resource-related problems and issues.

STAFFING QUESTIONS

What questions must the functional managers ask themselves before assigning resources?

Functional managers are required to staff several projects and must maintain a strong, well-trained, and efficient team of resources. Therefore, the functional manager must consider very carefully the answers to the questions identified below when developing his or her team:

- What are the requirements for an individual to become a successful team member?

- What skills are required to perform project-related work?

- What grade levels are needed for project assignments?

- Which individuals are qualified to be an assistant project manager?

- What problems can occur during recruiting activities?

- What projects are highest in priority?

- What problems or conflicts currently exist with the functional group?

Not all employees work well on project teams. Some people prefer assignments where they work by themselves. With regard to the third question, assigning a grade 8 to perform the work of a grade 6 may create a disgruntled employee. Assigning a grade 6 to perform the work of a grade 8 could be viewed as an opportunity for an immediate promotion. Functional managers should attempt to match work assignments with the appropriate grade and skill levels.

During and after recruiting activities, priorities can change that may result in mandating higher-grade employees to be assigned to other projects. This can be expected to happen on occasion, and project managers must be prepared if this situation occurs.

WORKER UNDERSTANDING AND SKILLS

Workers must have a complete understanding of why they were assigned to the project and what is expected of them.

Both the project manager and functional manager must make expectations clear to the assigned employees. Problems can occur when the project manager and functional manager are not in agreement regarding expectations. Typical items that assigned employees should be aware of include:

- Employees must know what work they are expected to perform, preferably in terms of an end product or deliverable.

- Employees must have a clear understanding of their authority and limitations with regard to the assignment.

- Employees must know what working relationships will be established or will be necessary during project planning and execution.

- Employees should know where and when they are not meeting expectations.

- Employees must be made aware of what can and should be done to correct unsatisfactory results.

- Employees must feel that both their functional manager and the project manager have an interest in them as individuals.

- Employees must feel that their functional manager and project manager believe in them and will support them as necessary to achieve success.

SPECIAL REQUIREMENTS

Are there any special situations or requirements that can impact the grade level of the assigned worker?

There are special situations that can influence the functional manager's decision on whom to assign. Examples might include:

- *Part-time versus full-time assignment.* Some functional employees find it difficult to work on several projects at the same time or to be assigned to different projects at different times each day.

- *Project manager is the functional manager.* When functional managers also act as project managers, they tend to focus on their specific needs and will keep the most qualified resources for their own projects.

- *Project manager is the general manager.* When this occurs, you can be sure that the best resources will be assigned to the project.

- *Several projects assigned to one project manager.* This may create some conflicts regarding priorities and the assignment of work to employees. The project manager can create an environment of pressure for the worker to produce deliverables for the assigned projects.

RECRUITMENT POLICY

Should there be a standardized recruitment policy for assigning employees to project teams?

Companies usually obtain the desired resources and function better when they have standardized recruitment policies.

- Unless some other condition is paramount, project recruiting policies should be as similar as possible to those normally used in the functional organization for assigning people to new jobs.

- Everyone assigned to a project should be given the same briefing. The information may be modified to accommodate different managerial levels, but everyone in the same general job classification should receive the same briefing. The briefing should be complete and accurate.

- Any commitments made to members of the team about treatment at the end of the project should be approved in advance by general management.

- Every individual selected for a project should receive an explanation about why he or she was chosen.

- Some flexibility should be granted to all people, or at least all those within a given job category, in the matter of accepting or declining a project assignment. (This depends on project needs and company policies.)

DEGREE OF PERMISSIVENESS

How much freedom should an employee be given to accept or decline an assignment?

There are several degrees of permissiveness within which functional managers may operate:

- The functional manager explains the project to the worker, and the worker is asked to join. The worker is given complete freedom to decline the assignment.

- The individual is told that they will be assigned to the project. However, they are invited to identify any reservations that they may have about joining. Compelling reasons may result in being excused from the assignment and this request could be associated with a personal or career preference.

- The individual is assigned to the project with no opportunity for discussion. Only an emergency would excuse the individual from serving on the project team.

THE PROJECT MANAGER'S RECRUITMENT CONCERNS

What issues and concerns affect the project manager rather than the functional manager during recruitment activities?

The project manager has recruitment/staffing issues similar to the functional managers. Some of these concerns include:

- Functional managers generally have the final decision regarding assignment of resources.

- Functional managers are often responsible for staffing multiple projects and may receive pressure from several project managers regarding assignment of resources.

- Resources are often unaware of the benefits of the project to the organization or how the project may benefit the resources individually.

- In some cases, promises or commitments are made to a resource during recruitment. These commitments should be documented and communicated to the managers involved. Plans should be developed to follow through on the commitment.

- Conflicts between project manager and functional managers are common during recruiting and staffing discussions. These conflicts should be addressed and resolved as quickly as possible. It is better for conflicts to be resolved during the initial planning stages than to experience major confrontations later.

MANAGEMENT PLAN DATA

EMPLOYEE: EXPERTISE:

EDWARD
WIGGINS

_____

_____   0   20   40   60   80   100

_____   PERCENT TIME ON PROJECT

_____

Note: A responsibility assignment matrix may accompany this data.

Sometimes during competitive bidding activities, a company is required to identify in its proposal what percentage of time the assigned employee will be working on the project. For example, in the illustration on the previous page, the company's proposal stated that Edward Wiggins would be assigned to the project and that he would spend 60 percent of his time on this project. In this case, the staffing of the project was committed to as part of the competitive bidding process and must be adhered to.

The difficulty with this approach is that the functional manager must be willing to commit the resource to the project very early in order to win the contract. There is no guarantee that the contract will be won and, even if the contract is won, the exact start date of the project may not be known. A responsibility assignment matrix (RAM) as shown earlier, along with the committed resource time assignment shown in this illustration, may accompany the proposal.

STAFFING PATTERN VERSUS TIME

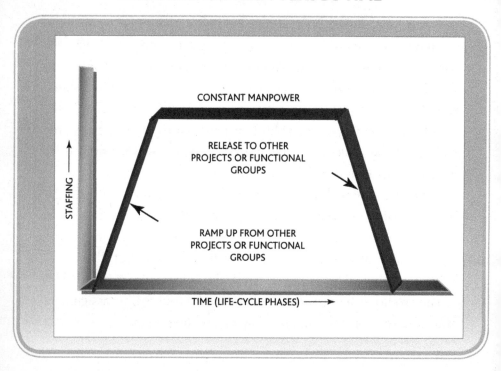

The illustration on the previous page shows typical manpower planning for a project. The ramp-up line indicates when staffing appears. Workers can be assigned from functional groups or transferred from other projects that have been completed. Ideally, project managers would like to have a constant manpower level on the project, with the same people remaining in place, rather than continuously releasing resources and having new people assigned.

Project managers must be willing to release their resources for other assignments as soon as their assigned work on the project is completed. This is referred to as destaffing a project. Project managers who hoard resources until the last minute are doing a disservice to the functional managers and other project managers, who need these critical resources. Destaffing a project is just as important as staffing the project and must be discussed in the project's human resource staffing plan.

SPECIAL ISSUES WITH ASSIGNMENTS

- Employees who view the assignment as a "chance for glory"
- Employees who work better by themselves than in a team environment

There are two major problems that functional managers must deal with, especially with highly talented employees. Some employees want a specific assignment and will employ the tactics necessary (including politics) to obtain the assignment because it is viewed as an opportunity and a chance for recognition or glory. These workers are often highly talented but may be more interested in exceeding the project's specifications rather than meeting objectives. Exceeding project specifications could lead to significant cost overruns. These people may also view this assignment as a necessary step in their career path and focus on the possibility of a near-term promotion.

Some employees simply do not work well in teams. Highly technical employees often do not agree with or trust another employee's ideas or solutions. They may insist on the use of their ideas or require validation of other ideas before reluctantly accepting them. When these people are placed in charge of small projects, they often perform all of the work themselves and delegate mundane work for their team members to perform. These employees are generally competent and very good workers, and it may be necessary to provide them with assignments on which can they work alone or with minimum contact with others. It should be understood that, although not the most desirable situation, not all employees are well suited to work in a team environment.

CONFLICTING POLICIES AND PROCEDURES

How should a line manager handle a situation where the project's policies and procedures conflict with the departmental policies and procedures?

Some projects may require specific policies and procedures that may conflict with functional and company policies and procedures. Functional managers must receive information about these policies as soon as possible to ensure the assignment of employees who have the capability to work under the specific conditions and to understand the risks associated with the project.

The greater problem is that sometimes these project-specific policies and procedures are not communicated or fully defined until well after the employee is assigned to the project. In these cases, the project manager and all affected functional managers must work together to identify and either eliminate or minimize the potential problems associated with these special or unique policies and procedures.

ASKING FOR A REFERENCE

Should the executive responsible for selecting the project manager allow functional managers to provide an input on who should be assigned?

Not all project managers have equal skills when it comes to working with large project teams. Some project managers prefer small projects that require small teams of just two or three members. The main reason for this is leadership capability. Project managers may find it necessary to develop and engage in different leadership styles for each functional employee assigned. In some cases, the demands of leadership are too great for the project manager. Project managers who lack experience or have not developed strong leadership skills tend to perform poorly on large projects where decisiveness and the ability to motivate are essential.

Some functional managers or groups seem to work better with certain project managers, especially if the project manager is familiar with the problems and issues that the functional manager must cope with on a daily basis. It is an acceptable practice for project sponsors to ask the functional managers to identify which project managers they prefer to work with on large projects, especially if a significant number of resources from one functional department will be assigned to the project.

A SUMMARY OF OTHER SPECIAL ISSUES

What are some of the special issues facing functional managers during project staffing activities?

The functional manager's job can be a real challenge when it comes to project staffing. Following is a summary of some of the issues:

- If the project manager finds that the assigned resources are not acceptable, the problem may be escalated to higher-level managers, and there is a possibility that the project manager may outsource the work.

- On a long-term project and the possibility of full-time assignments, the functional manager may be concerned with how to provide the employee with career development opportunities, especially if the resource is located some distance from the functional manager.

- How will employee performance appraisals be prepared and delivered?

- Some employees share their time between multiple projects and mandatory functional work. Which employees will perform well in this situation, especially if the functional work has a higher priority that project-related work?

- Will the assignment fall within the employee's current rank and salary?

- There may be situations in which the functional manager may be forced to reassign resources committed to a project due to a higher priority.

- Multiple projects may deplete the available resources.

THE FUNCTIONAL MANAGER'S PROBLEMS

Unlimited work requests (especially during competitive activities)

We can now summarize some of the significant problems facing the functional manager.

Companies that survive on competitive bidding must assume they have unlimited resources when bidding on contracts. These companies do not know how many contracts they will win, if any, but must assume that they will have sufficient resources to support any successful bids.

This creates havoc for functional managers, who may find their resources overcommitted on several projects. Predicting resource requirements in this type of competitive environment is extremely difficult.

Hiring additional resources may also present some challenges. Training may be required, and some training may be performed by the overcommitted resources, creating additional problems. A lack of experience and lack of understanding of company and project policies could cause other issues, and there is a need to transition new employees into the existing functional group, which may be difficult at times.

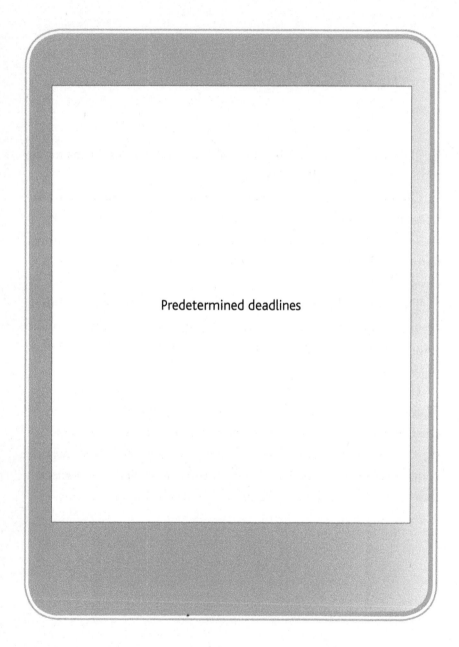

Predetermined deadlines

During competitive bidding activities, predetermined deadlines are often negotiated. It is then up to the contractor to determine the start date to meet the predetermined deadlines. Therefore, although not recommended, some project plans are prepared from the end date back rather than from the start date.

Predetermined deadlines have a combination of positive and negative effects for functional managers. While the deadlines indicate when the resources will be available for assignment to other projects, the amount of time the resources must remain committed to a given project is unknown.

Predetermined deadlines can change. Projects may undergo scope changes that can alter project schedules. Deadlines can also change due to external project dependencies or due to work associated with other projects that may delay resources needed to meet contractor commitments to your project.

All requests having a high priority

Perhaps the single most important item to consider during project staffing is the project's priority relative to other projects. Unfortunately, it is common for every sponsor and every project manager to believe that his or her project is the highest-priority project within the company regardless of the priorities that were established within the project portfolio.

Typical rationalization for establishing priorities includes:

- The project's technical risks

- The company's risks, financially and competitively

- The urgency associated with the deliverables and the end date of the project

- The expected savings, profitability, or return on investment (ROI)

- Future contracts with the customer

- The impact of the project on other projects

- The connection to organizational business goals

Functional managers may not always follow the priorities when assigning resources. For example, the best available functional resource may be the only person available to perform the work required on a lower-priority project. Also, if the best resource is already working full time on a project, it could be a mistake to change resources and disrupt progress. Priorities may be viewed as guidelines for functional managers, not mandates.

Limited number of resources

For companies that survive on competitive bidding, it is assumed during the bidding process that the company will have the resources needed to perform the work. Companies often bid on more projects than they can handle at one time, knowing that it is unlikely that they will win all of the contracts they are bidding on.

In reality, functional managers have a limited number of resources and may not be able to identify and commit a specific resource for each project. The best that the functional manager may be able to do is a commitment for a specific skill level or pay grade.

Resources are considered by the negotiators to be unlimited during the competitive bidding process, but the resources become limited when the contract is awarded and the calendar dates are fixed. Simply stated, contracted calendar dates limit resource availability.

When resources become limited, priorities become a factor when assigning resources. Project management offices and project portfolio management can help to resolve or manage resource issues more effectively.

Limited availability of resources

Resource availability may become a more important factor than the established priority of the project. Positioning a project as a high priority is no guarantee that the required resources will be available. As stated previously, functional managers are reluctant to remove resources from an ongoing project and cause a potential schedule slippage to assign the resource to a higher-priority project. Alternatives should be identified and risks assessed before making this type of decision.

Even when the functional manager performs excellent planning for resources, problems can occur. For example, Project A was expected to be completed in March, and a specific resource on Project A was scheduled for Project B in April. However, due to technical difficulties, Project A slips until July and the resource needed for Project B will not be available until July.

In this case, because of limited availability of resources, both Projects A and B will slip. Removing the resource from Project A to maintain the schedule for Project B may be an option, but the damage to Project A could be significant and result in a much further delay and other potentially serious business risks.

Unscheduled changes in the project plan

As stated previously in this book, projects are unique activities that may never have been done before and may never be done again. As a result, the estimates of time and cost developed during project planning may be very inaccurate. The result could be a major change in the project plan. These changes may involve the scope of work, quality of work, activity durations, and cost.

Very few projects are completed according to the original plan and established triple constraint. Trade-offs will probably be required, and the decisions about the order of the trade-offs are made jointly among the project manager, affected functional managers, and the project sponsor. Trade-offs and other such issues cause changes in the project plan, which in turn may affect the length of assignments of the resources committed to the project.

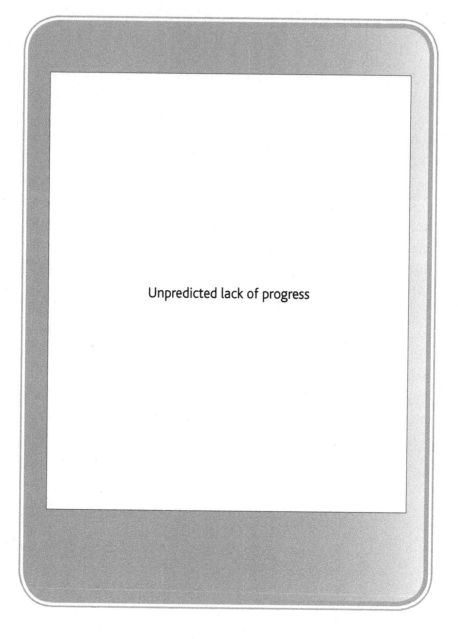

Project estimates imply an anticipated rate of progress. If the estimates are wrong or if problems occur, the project manager and team could experience an unexpected lack of progress. This can affect the release of critical resources and can impact downstream projects that are dependent on the release of resources from the project.

The unpredicted lack of progress could be the result of unrealistic or optimistic original estimating. Technically oriented functional managers generally provide optimistic estimates. Other factors that cause a lack of progress include:

- Resources being assigned to perform work above their skill level and pay grade

- Resources being assigned to perform work below their skill level and pay grade, resulting in a lack of motivation and apathy

- Resources assigned to multiple projects, causing significant delays due to ramp up and lack of work continuity

- Resources being assigned to multiple projects and spending more time than necessary on the projects that interest them or motivate them the most

- Resources were never provided with clear expectations about the rate of progress required

- Resources did not clearly understand what they were expected to do

Unplanned absence of resources

An unplanned absence of resources is a common headache for both project managers and functional managers and may result in the shifting of resources from one project to another. If a functional resource is committed to several projects and must spend additional time on one project because of a crisis, all of the other projects can be adversely affected. The unplanned absence could be the result of:

- Resources temporarily committed to other project to resolve a crisis

- Resources temporarily brought back to their functional area to resolve a crisis

- Functional work having a higher priority than project work

- Acts of God, such as bad weather

- Unplanned service disruptions—loss of power and other facility issues

- Labor and union issues such as strikes

- Vacations

- Religious holidays that the local labor force observes

Overtime is not always the solution to an unplanned absence of resources. Overtime can lead to mistakes and/or employee burnout if it is prolonged. The best solution is for functional managers and project managers to work together to develop contingency plans.

Unplanned breakdown of resources

The greatest amount of capital equipment in an organization is usually found in the manufacturing industry. The unplanned breakdown of one piece of capital equipment can result in a work stoppage. Companies try to minimize possible breakdowns in the nonhuman resources with routine maintenance and planned plant closings, but this does not cover all possible problems. Bad weather, acts of God, poor or incomplete maintenance, labor stoppages, and unpredicted equipment problems can negatively impact projects and the organization's work operations as well.

Unplanned loss of resources

Project managers would like nothing better than to hold on to their critical resources for the duration of the project. This is unlikely to happen because:

- The longer the project, the less likely it is that the project manager can retain the high-quality resources.

- The higher the quality of the resources, the less likely it is that they can be retained over the length of the project.

- Changes in priorities can cause resources to be reassigned to other projects.

- The higher the quality of the assigned resources, the greater the likelihood that they will be promoted and possibly removed from your project.

- Projects undergo the problems that companies do with regard to turnover of personnel due to retirement, promotions, and leaving the company for employment elsewhere.

Some project managers prefer to have average or competent employees assigned to their project rather than the best workers. The reason is that project managers may prefer to have the same faces on their project from beginning to end rather than to fight for the best resources and be unsure as to the ability to retain them over the project's duration. There is a greater probability that average or above-average people can be retained, and the project manager has a greater opportunity to work with and develop the skills of the employees.

THE FUNCTIONAL MANAGER AS A FORECASTER

- Unable to determine which resources will be available for a future go-ahead date
- Also, unable to determine the future go-ahead date

One of the biggest challenges facing the functional managers is attempting to predict the future. The complexities include the following:

- The functional managers must be able to determine which resources will be available at a future date. Some companies require that functional managers maintain a three- or six-month availability window for all employees.

- The go-ahead date on many projects is not well defined. The date can change because of the demands of other, higher-priority projects; the company's predicted future cash flow; and the company's anticipated profitability.

- The functional manager must be able to foresee the need for new hirers, how long it will take to train them, and who will perform the training.

- The functional manager must attempt to predict what problems could exist in the future that may affect resource availability and impact the assignment of resources.

Some companies do a good job using portfolio management to lay out a plan for the start date of selected projects. Unfortunately, priorities change as new projects are added to the queue, and other projects may no longer be considered necessary. The problem is further complicated in companies that survive on competitive bidding, where the number of accepted bids is extremely difficult to predict.

THE TYPE OF MATRIX STRUCTURE

There is a relationship between the type of matrix desired by either the project manager or functional manager and the quality of the resources assigned.

Most companies today have some form of matrix structure. There are three basic types of matrix structures, and the effectiveness of each is based on who has greater influence over the workers on a daily basis. Typical matrix structures are:

- *Strong matrix.* In a strong matrix, the project manager has a greater influence over the workers than the functional manager has. In this type of structure, the functional manager may assign high-quality resources and allow them to receive daily direction from the project manager, with minimal interfacing with their functional manager.

- *Weak matrix.* In a weak matrix, the functional manager has a greater influence over the workers than the project manager. The functional manager may assign less qualified or average resources to the project, knowing that they are under the direct supervision of the functional manager. The functional manager works very closely with these employees on a daily basis.

- *Balanced matrix.* In this matrix form, the resources assigned to a project may be a combination of average to highly qualified and take direction from both the project manager and their functional managers. The most common issue in this type of structure is referred to as the "two boss syndrome".

It should be understood that the functional manager has significant influence regarding the type of matrix structure that will be used. Generally, the resources are "owned" by the functional groups. In some cases, part of the project can be managed as a strong matrix and part can be managed as a weak matrix, both occurring at the same time. This depends on the technology and complexity of the project.

THE FUNCTIONAL MANAGER'S VIEW

How do functional managers view the role of the project manager?

Even though most functional managers may have a basic understanding of the role of the project manager, there are still different perceptions of the project manager's role. The perception of the functional manager is often based on whether the company is project-driven or non-project-driven. In project-driven companies, the functional manager's view might be:

- The project manager is simply the customer's contact point, and the real work is done by the functional managers.

- The project manager is the company's profit center, whereas the functional areas are treated as cost centers.

- Project managers share in bonuses and rewards, whereas functional managers may not share in the recognition for a job well done.

- Project management is the stepping-stone to promotion to a senior management position.

In non-project-driven companies, the view might be:

- The project manager is simply a coordinator or expeditor of work.

- Project managers are people who have been forced to manage a project and cannot wait to get the job done so they can return to their former functional position.

- Project managers are just paper pushers.

- Project managers perform work that functional managers dislike doing.

WORKING WITH THE PROJECT MANAGERS

What is the working relationship between the project manager and the functional manager?

What expectations does each have of the other?

The working relationship between the project and functional manager must be a collaborative working relationship. Conflicts between the project and functional managers must be resolved quickly and, when necessary, with project sponsor involvement.

Functional managers expect the project managers to:

■ Provide a clear description of the work that must be accomplished by their functional department.

■ Provide a clear explanation of all project deliverables.

■ Provide information that will assist in developing a realistic timetable for accomplishing the functional work.

■ Identify special requirements in other functional areas that can impact a functional manager.

■ Understand that problems will occur resulting in the loss or change of resources.

■ Work with the functional managers in developing contingency plans to address project risk situations and unexpected problems.

Project managers expect the functional managers to:

■ Assign the correct resources.

■ Meet all of the deliverables according to the timetable.

■ Provide reliable estimates and assume accountability for the estimates.

■ Demonstrate a willingness to work with the project manager in resolving problems.

EXPECTATIONS OF THE ASSIGNED RESOURCES

What expectations do the project and functional managers have of the assigned resources?

The expectations commonly communicated to the assigned functional resources include:

- Accept responsibility for accomplishment of the assigned deliverables within the imposed constraints. This implies that the assigned resources must agree to effort and duration estimates even if the estimates were made by someone else.

- Complete the work at the earliest possible time. The assigned resources must be willing to compress rather than expand work when possible.

- Periodically inform both the functional manager and project manager of the project's status. This includes the real project status such as hours spent, work accomplished, and time and money needed to complete the assigned work package.

- Bring problems to the attention of the project manager and functional manager quickly for resolution. Team members must realize that the project and functional managers are there to help team members resolve issues. People should not be reprimanded for communicating problems.

- Share information with the rest of the project team in a timely manner.

- Demonstrate a willingness to accept feedback and constructive criticism.

HANDLING ORGANIZATION PRIORITIES

Can the priorities of the ongoing business be more important than priorities among the projects?

In both project-driven and non-project-driven companies, there are situations where priorities are associated with the ongoing business rather than on the projects in progress. Typical reasons include:

- A large portion of company profits and revenue, in many cases, comes from operations and the functional units rather than from projects.

- Many projects exist to support activities and programs in the functional units.

- Projects may have anticipated profitability attached, but there may be crises in the functional units that could seriously jeopardize the ability to conduct business.

- The future of the company rests with the capabilities of certain functional units, even though some of the projects have expected profit margins.

HANDLING PROJECT-RELATED PRIORITIES

How do functional managers handle priorities, especially priorities among staffing projects?

Priorities are usually the driving force for assigning resources. But is it the company's established priority list or the functional manager's personal priority list? The company's priority list simply states which projects are more important than others from a business perspective. It is the functional manager's perception of priority that determines which resources will be assigned. Examples might include:

- The manager of the research and development (R&D) department has to staff 10 high-priority projects but keeps his best resource committed to a departmental project that could improve the company's capability in the future.

- The functional manager assigns average or above-average resources to the high-priority projects, but the best resources are assigned to departmental-related work, where they would be on call as consultants to support any project that may experience problems.

- The functional manager believes that some of his or her best resources may be promoted to a position where they would no longer be able to work on projects. It would be potentially damaging to remove them from projects at an inopportune time.

- The functional manager has high-quality resources available, but they do not work well with others on teams.

BALANCING WORKLOADS

How does a functional manager balance operational work with project work?

Previously, we stated that in project-driven companies the functional units are cost centers and exist to support the projects. In non-project-driven companies, the functional units are profit centers and the projects exist to support the operational work in the functional units.

Regardless of how the company is driven, the functional manager must balance ongoing work with project work. Factors that can influence the functional manager's ability to balance work include:

- The importance and consistency of the company's priority list (continuous changes in priorities)

- The importance of the functional manager's personal priority list and the influencing skills of the functional manager

- The quality of the available resources

- The quality of the resources already working on projects

- The quality of the resources performing operational work

- Ability to subcontract out operational work

- Ability to outsource project work

- Ability to hire full-time or temporary employees

- Quality of the resources that are hired

- Strategic importance of selected projects

- Personal whims of management and sponsors

MULTIPROJECT PLANNING

How do functional managers effectively handle multiproject planning activities?

How far out should the planning window be?

What factors can affect the length of the planning window?

One of the complexities of the functional manager's job is that they must perform multiproject planning rather than planning for one project at a time. Most functional managers use a planning window, which can be three, four, or possibly six months in duration. The factors that can influence the length of the planning window include:

- Quality of the functional estimates for duration and effort

- Identified project risks

- History of scope changes on projects

- Quality of available resources

- Duration of the projects

- The portfolio priority list

- The existence of a corporate capacity planning model

CHANGING RESOURCES DURING THE PROJECT

How does a functional manager decide which resources can be replaced with other, higher- or lower-quality resources during the project?

What variables affect the decision?

What are the potential ramifications of changing resources?

The higher the quality of the assigned resources, the greater the likelihood that they will not remain on the project through its entirety. Resources can be expected to change over the duration of the project. Reasons for changing from higher- to lower-quality resources include:

- Most of the critical work has been completed, and lower-quality workers can adequately complete the remaining work.

- A critical resource is needed on another project or to handle a crisis in the operational area.

- When the project started up, only higher-quality resources were available.

There are also valid reasons for assigning a higher-quality resource:

- The current resources cannot perform the work in a satisfactory manner.

- The project's schedule must be accelerated.

- The customer is demanding higher-quality resources.

- The current resources cannot cope with the project's demands and risks.

- Project team members have been promoted to other positions.

- Project team members have resigned from the company.

THE IMPACT OF SCOPE CHANGES

How important are freeze dates?

How important is it to follow the change control process?

Freeze dates are imposed dates designed to prevent further configuration changes or scope changes. Milestones may accompany or may be negotiated in connection with freeze dates. The establishment of freeze dates may be based on other activities in the customer's organization or the need to move forward with technical work and production of deliverables. Freeze dates may be needed to prevent slippages of other project activities.

Freeze dates can change provided that a change control process is in place and followed. Scope changes can change the freeze dates or extend the freeze period. This is one of the reasons why a change control board is established. The change control board is usually comprised of key stakeholders and/or decisions makers from both the customer and the contractor organizations. At the change control board meetings, the functional manager and the functional team member requesting the change will assist the project manager in presenting the rationalization for the change. At the meeting, four critical questions must be addressed:

- What is cost of the change?

- What is the impact on the project schedule and freeze dates?

- What value-added or additional benefits will be realized as a result of the change?

- What are the risks to the project, other projects, and the organization?

If the customer or sponsor agrees to the scope change, the freeze dates can be moved. If the freeze dates cannot be moved due to the implications associated with the change, the scope change may not be approved, even though there may be considerable benefits for the customer. Changes should be reviewed as projects. There are risks, benefits, and costs associated with any change, and in some cases a cost-benefit analysis may be required.

RISK MANAGEMENT

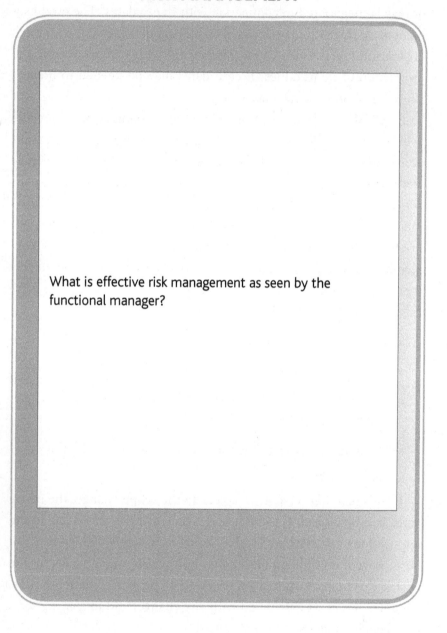

What is effective risk management as seen by the functional manager?

Functional managers generally accept accountability for the deliverables created by their functional organization and therefore focus heavily on functional risk management rather than project risk management. As such, functional managers may have developed their own templates for risk management. However, functional managers generally have a narrow focus regarding risk and look only at selected risks and proven responses:

- *Risks in the estimates for duration and effort.* Mitigation strategy is to use more highly skilled labor.

- *Risks in the estimates of cost.* Mitigation strategy is using above-average labor that can do the job in less time.

- *Risks in technology.* Mitigation strategy is to maintain a technically competent labor force.

Other risks:

- *Business-related risks.* Mitigation is the responsibility of the project manager and the project sponsor.

- *Integration risks.* Mitigation strategy is the responsibility of the project manager.

PROJECT DOCUMENTATION

How does a functional manager determine the appropriate amount of documentation for project-related work?

Is the decision made by the functional manager or the project manager?

Can the paperwork requirements be dictated by the enterprise project management (EPM) methodology?

Paperwork is generally a headache for both the functional manager and the project manager. Functional managers are interested more so in the paperwork related to the technology in their group and any information that can be used to improve the accuracy of the estimates they must provide. The project manager, however, wants standardized reporting that every functional manager can use.

Project managers generally make the final decision on project-related paperwork, but input from the functional managers is recommended. If the functional manager does not support the project manager's decision, then the functional manager may negotiate for the right to create his or her own paperwork and documentation requirements for the workers to follow.

Some companies have enterprise project management (EPM) methodologies that dictate the documentation requirements for all employees to follow. The paperwork and documentation requirements of the EPM system are based on the needs of:

- The project team

- Senior management

- Functional management

- Customer requirements

- Regulatory agencies, if applicable

CONFLICTS

What are the most common conflicts between functional managers and project managers?

There are numerous conflicts that can exist between the project manager and the functional manager. A few of the most common conflicts involve the quality of the assigned resources, activity duration, and cost estimates; and personality issues among the project manager, functional manager, and the assigned resources.

Project managers often demand the best resources from the functional managers. While there are downside risks associated with the acquisition and retention of these resources, project managers will (understandably) demand the best. Functional managers know the capability of the resources better than the project manager even if the project manager has had previous working relationships with these resources and make their assignment decisions based on the needs of all projects for which the functional manager is responsible.

From a project manager's perspective, one of the most serious conflict situations is personality clashes between team members. If the assigned resources, regardless of their capability, have the potential for serious personality clashes with other team members, it may be better for the project manager to have other people assigned. Some project managers will interview each potential team member, in a one-on-one session, to determine if they will be compatible with the other team members. The recommended time to identify potential personality conflicts is in the staffing process for the project rather than during project execution.

CONFLICT RESOLUTION

What is an effective way for the functional manager to handle conflicts with project managers?

Is there an escalation process that can help with the resolution process?

In the previous pages, we discussed conflicts over manpower resources and personalities. There are other conflicts that can arise, such as disagreements over cost estimates and incurred costs, time estimates and delays in activity completion, quality, priorities, technical opinion, trade-offs, equipment and facilities, administrative procedures, and responsibilities.

Project and functional managers must work together on a continuous basis, and any conflicts that arise must be resolved through confrontation and collaboration. Each party in the conflict must be willing to see the conflict through the eyes of the other person and must be willing to work together to develop a solution as quickly as possible.

It is preferable for conflicts to be resolved at the project-functional interface level. But not all conflicts can be resolved by the project manager and functional manager, and the project/executive sponsor may be required to step in and assist with the resolution of the conflict. The result in this situation may or may not be favorable to one side, and there may be a winner and a loser. Each person must understand that the outcome of a conflict will not always be decided in his or her favor and the decision should not jeopardize the working relationship between the project and functional managers. Conflicts should not be perceived a negative situation. Conflicts can produce favorable results for both sides and the organization as a whole if they are managed correctly.

TALKING TO PROJECT MANAGERS

How often should functional managers and project managers converse?

What factors affect the frequency?

Functional managers need to know both the status of the project and how well or poorly their employees are performing. Functional managers do not have the luxury of being able to attend each and every project team meeting because of time constraints and the demands of multiple projects or operations work. As a result, functional managers must be briefed by both their employees and the project manager.

Functional managers generally prefer to hear information from the project manager rather than the assigned workers. Workers sometimes withhold information from their functional manager, fearing that this information could be used against them during performance reviews. Another reason for seeking feedback directly from project managers is that the project manager often informs the functional manager of information that may be withheld from the team for reasons of confidentiality.

As a rule, it is a good idea for the project manager to communicate with each functional manager at least once a week. If the project is considered critical or is highly visible and the functional manager's group is actively involved in project work, then communication may be more frequent.

PROJECT PERFORMANCE REPORTS

What are the types of project performance reports?

What input is required by the functional manager?

How is the input provided?

S tatus reports are needed to provide information to all stakeholders about the condition of the project and what progress has been made. There are three types of status reports:

- *Progress reports.* How much work has been accomplished, and what did it cost us to do this work?

- *Status reports.* What is the current project condition, and how does the work that has been accomplished compare with the baselines for time and cost? In other words, are the variances favorable or unfavorable?

- *Forecast reports.* Based on level of performance to date, what is the estimated final cost and anticipated completion date for the project?

All of these reports are computerized and generally part of the company's EPM system. There are two inputs required by the functional managers:

- How much work has been completed during this reporting period?

- What was the cost to complete the work?

To determine how much work was completed, the functional manager must provide input about the percentage of work completed or by a formula by which the amount of work earned can be calculated. The actual costs associated with the work may come from account codes connected to employee time reports and accounts payable for equipment and other resources.

The accuracy of the information in the performance reports is highly dependent on the functional manager's input.

ESTIMATING AND SCHEDULING

What difficulties do functional managers face with regard to estimating and scheduling?

Are the difficulties related to the quality of the estimates?

In the previous pages, we discussed that variances in performance are determined by comparing actual performance to baselines. The baselines were developed for the most part using estimates provided by the functional managers. The quality of the estimate is based on the functional manager's experience, judgment, and estimating method. As an example, the functional manager has collected data over the years and concluded that the standard to complete a certain work package is 500 hours using a grade 6 employee. Functional managers can use this standard to develop other estimates that are based on the resources that will be used:

- If the assigned employee is a grade 5, the hours will be increased.

- If the assigned employee is a grade 7, the hours will be reduced.

- If the employee is a grade 6 or higher but has never performed this activity previously, the hours may be increased.

- If the functional manager believes that this work package is technically 10 percent more difficult than the standard, then the hours will be increased by 10 percent.

Just from this example alone, it should be obvious as to the complexity of estimating. Now, imagine the problems facing the functional manager if the standard did not exist. The estimate would be considered a "seat-of-the-pants estimate" (a wild guess) and highly unreliable.

AN EFFECTIVE WORKING RELATIONSHIP

What is an effective working relationship between the functional managers and the project managers?

- Working with the right project managers
- Working with the wrong project managers

Previously, we stated that the executive sponsor may ask the functional managers for input regarding which project manager they prefer to work with. Unfortunately, functional managers do not often have this luxury and must work with whoever is assigned. When the functional manager works with the right project managers:

- The working relationship is collaborative.

- The project manager briefs the functional manager periodically on the project's status and the performance of the assigned workers.

- The majority of the conflicts that occur are resolved at the project manager–functional manager interface.

- Contingency plans are developed jointly.

- Information is exchanged freely and efficiently.

When working with the "wrong" project managers:

- The working relationship becomes confrontational and combative.

- The functional managers are briefed by their workers, with little interfacing with the project managers.

- The majority of the conflicts are resolved by the project sponsor or senior management.

- Each party develops his or her own contingency plans without communicating information.

- Information is exchanged only when mandated.

SUCCESSFUL CULTURE

What are the elements of a successful culture to support the project manager–functional manager interface?

While it is true that the project manager and functional manager can develop a mutually agreeable project culture and working relationship, the corporate culture may dictate the actual relationship. There are four typical cultures:

- *Cooperative cultures.* The relationship between the project manager and functional manager is based on trust, communication, teamwork, and cooperation.

- *Competitive cultures.* Each party tries to advance at the expense of the other party.

- *Isolated cultures.* The functional unit creates its own culture, and the project manager must manage work according to that culture or risk alienating the functional manager and the functional group.

- *Fragmented cultures.* These appear on multinational projects where the project manager must coordinate work globally and through virtual teams.

The most desirable culture is obviously the cooperative culture. When it exists:

- An effective and well-defined working relationship is established between the project manager and those line managers who directly assign resources to projects.

- The functional employees working on projects find it much easier to report vertically to their line manager at the same time they report horizontally or dotted line to one or more project managers.

PROMISES MADE

What promises can a project manager make to functional team members with regard to:

- Promotion
- Salary
- Bonus
- Overtime
- Responsibility
- Future work assignments

Unless the project manager and the functional manager are the same person, project managers have virtually no responsibility for wage and salary administration. Yet some project managers try to motivate the team by making promises that cannot be filled. This may create significant interpersonal conflict. Consider the following:

- Project managers cannot promise a functional employee a promotion. The project manager may be allowed to provide a recommendation to the functional manager, who will make the final decision.

- Project managers cannot promise employees salary increases. This is a line function.

- Project managers may have money allocated in their budget for bonuses, but may need the functional manager's approval.

- Project managers may have funding available for overtime on the project, but the final decision resides with the functional managers.

- Project managers cannot allow workers to perform work above their pay grade unless approved by the functional managers (especially in collective bargaining or union situations).

- Project managers cannot promise employees future work assignments. Project managers can request the worker to be assigned to future work, but cannot promise the assignment.

NON-FINANCIAL AWARDS/RECOGNITION

With non-financial awards, employees may receive cash-equivalent items, but not cash-in-hand.

Project managers generally have no direct responsibility for wage and salary administration. This responsibility depends on the type of organization and organizational structure. If the project manager is also the functional manager, there may be some responsibility regarding wages and salary. Therefore, project managers, generally speaking, cannot provide direct cash awards to team members other than perhaps bonuses that have been included in the project's budget and are approved by the sponsor.

Project managers may be in a position to offer noncash recognition awards, and this can be done without approval of the functional managers or sponsor. Typical noncash awards include:

- Theater tickets

- Tickets to athletic events

- Certificates for fine dining

- Use of the company car

- Management-granted time off

- Plaques, newsletter articles, or other recognition methods

- Gifts from a catalog

- Paid vacation

These types of noncash awards can be provided by the functional managers as well but are more commonly given out by project managers.

WALL-MOUNTED PLAQUES FOR ALL TO SEE (CAFETERIA WALL)

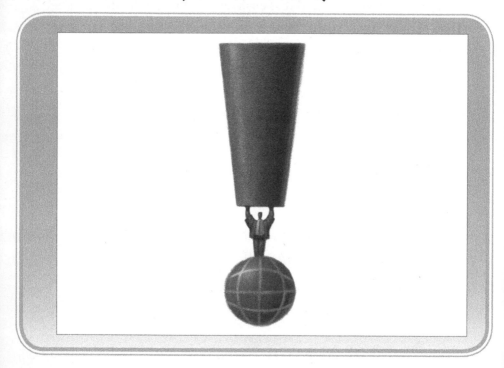

Sometimes the recognition associated with an accomplishment means more to the worker than receiving cash or something of monetary value. As an example, one company institutionalized a "Wall of Fame" that could be seen as the employees entered the company cafeteria. Whenever an employee did something outstanding for the company or the project, recognition was provided in the form of a plaque and mounted on the wall.

Another form of recognition is to publish an article about the employee and acknowledge their accomplishments in the company newsletter.

Another company provided certain employees with wooden tokens, similar to Boy Scout merit badges, for performance well done. The tokens were good for a free lunch or snack in the company's cafeteria. What the company found was that the employees were performing exceptionally well trying to collect as many tokens as possible. The tokens were mounted in each employee's office for all to see. The motivation was to see how many you could collect.

PUBLIC RECOGNITION

As mentioned previously, the recognition itself is often more important to the worker than the actual award or any dollar value that comes with the award. Consider the following two examples:

- A company-initiated a policy that any employee who becomes a Project Management Professional® (PMP) would receive a one-time bonus of $500. The company did not acknowledge publicly who became a PMP, but the next paycheck for the employee contained the bonus. The company then changed their policy, and once a week, in the cafeteria at lunch, an executive of the company would publicly recognize all of the people who had become PMPs during the past week. The bonus was still provided, but more people were now taking the exam because of the public recognition.

- Employees who performed well were not only recognized publicly in this company but were provided with elegant plaques ready for mounting. The employees mounted their plaques in their offices for everyone to see rather than taking them home.

In both cases, the public recognition was more important to the workers than the award.

OTHER NON-MONETARY AWARDS

There are other examples of non-monetary awards:

- A company maintained a fleet of cars for the sales force and some of the executives. However, employees whose performance on projects was considered outstanding would receive use of the company car for a week or two. This was considered a bonus by the employees who had to commute large distances each day.

- A company maintained a box at certain sporting events and at certain theater events. Employees were given access to these seats for a job well done.

- A company completed a three-year project, and the customer was elated. Everyone in the company knew that the success of the project was due to one blue-collar union member who worked excessive overtime to make the project a success. Unable to reward him financially, the company gave him use of the company credit card for a week-long paid vacation for him and his family. The union commended the president for recognizing the contributions of the employee toward the success of the company.

- A company had a relationship with a very elegant restaurant where the company would entertain clients. Employees who performed well were given three free meals over a period of a month.

PUBLIC PAT ON THE BACK

Public recognition does not necessarily require a formal reward-and-recognition event or an award. A simple "pat on the back" or other expression of thanks and appreciation can suffice. However, there are some people who fail to use reward and recognition properly or abuse the concept and associated processes. Consider the following:

- A project manager believed that people should be told that they are doing a good job. This belief was based on a desire and a need to motivate the workers. The problem in this case was that the project manager was providing positive feedback even for an employee who was performing poorly in an attempt to motivate the employee toward better performance. When the employee received a below-average performance review, he argued that the project manager indicated on several occasions that he was doing a great job.

- Rather than recognize everyone, a project manager believed that no one should be told they are doing a good job if they are merely doing the job they were expected to do. The project manager recognized the performance of only those people who went above and beyond what was expected of them. This led some people to believe that the project manager was displeased with their performance.

SECURING PROPRIETARY KNOWLEDGE

One of the most critical issues facing companies today is how to safeguard proprietary information. In most cases, it is the responsibility of the functional business unit and functional manager to provide instructions, policies and procedures about how information should be protected.

Many times, customers will talk directly to the assigned team members without involving the project manager. In these situations, the project manager may not be aware of the information the employee shared with the customer. Even though companies have policies regarding information distribution, the responsibility for protecting information is generally assigned to the functional managers.

WEARING MULTIPLE HATS

We have been discussing throughout this book project situations where the functional manager and project manager are not the same person. In small companies with limited resources and in large companies where special projects have been initiated, the functional managers must wear multiple hats and act as the project manager as well as the technical expert or functional manager at the same time. In such cases where the functional manager also becomes the project manager:

- The functional manager has full wage and salary responsibility.

- The functional manager may retain the best available resources for his or her project.

- The functional manager has full authority over the assigned resources from his or her functional area.

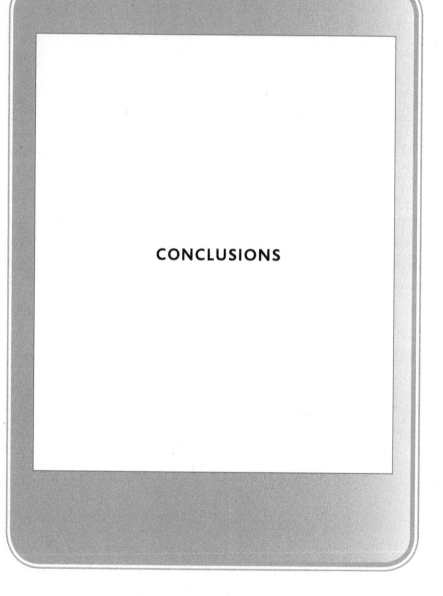

CONCLUSIONS

Functional managers have the power to drive a project to success or point the project in the direction of failure. The working relationship between the project and functional managers is important. Today, the authority and responsibility for project success is shared between the project and functional managers, rather than single-person total accountability in the hands of the project manager.

INDEX